BIG IDEAS
MATH®
Advanced 1

Record and Practice Journal

- Fair Game Review Worksheets

- Activity Recording Journal

- Practice Worksheets

- Glossary

- Activity Manipulatives

BIG IDEAS LEARNING®

Erie, Pennsylvania

About the Record and Practice Journal

Fair Game Review

The Fair Game Review corresponds to the Pupil Edition Chapter Opener. Here you have the opportunity to practice prior skills necessary to move forward.

Activity Recording Journal

The Activity pages correspond to the Activity in the Pupil Edition. Here you have room to show your work and record your answers.

Practice Worksheets

Each section of the Pupil Edition has an additional Practice page with room for you to show your work and record your answers.

Glossary

This student-friendly glossary is designed to be a reference for key vocabulary, properties, and mathematical terms. Several of the entries include a short example to aid your understanding of important concepts

Activity Manipulatives

Manipulatives needed for the activities are included in the back of the Record and Practice Journal.

Big Ideas Learning and Big Ideas Math are registered trademarks of Larson Texts, Inc.

ISBN 13: 978-1-60840-528-2
ISBN 10: 1-60840-528-1

89-VLP-17 16 15

Contents

Contents

Contents

Contents

Contents

Contents

Contents

Contents

Contents

Chapter 1   Fair Game Review

Determine whether the number is prime or composite.

1. 4

2. 7

3. 13

4. 22

5. 19

6. 27

7. 30

8. 37

9. 41

10. 45

11. You have 33 marbles. Besides 1 group of 33 marbles, is it possible to divide the marbles into groups with the same number of marbles with no marbles left over?

12. You have 43 pencils. Besides 1 group of 43 pencils, is it possible to divide the pencils into groups with the same number of pencils with no pencils left over?

Chapter 1 **Fair Game Review** (continued)

Add or subtract.

13. $1\dfrac{1}{5} + 1\dfrac{3}{5}$

14. $2\dfrac{3}{7} + 3\dfrac{2}{7}$

15. $4\dfrac{5}{9} + 6\dfrac{2}{9}$

16. $3\dfrac{6}{11} + 5\dfrac{4}{11}$

17. $4\dfrac{3}{4} - 2\dfrac{1}{4}$

18. $5\dfrac{3}{8} - 3\dfrac{7}{8}$

19. $2\dfrac{3}{10} - 1\dfrac{7}{10}$

20. $6\dfrac{5}{12} - 2\dfrac{11}{12}$

21. You are baking cookies. You have $7\dfrac{1}{4}$ cups of flour. You use $2\dfrac{3}{4}$ cups of flour. How much flour do you have left?

Name_____ Date_____

1.1 Whole Number Operations
For use with Activity 1.1

Essential Question How do you know which operation to choose when solving a real-life problem?

---

1 ACTIVITY: Choosing an Operation

Work with a partner. The double bar graph shows the history of a citywide cleanup day.

- **Underline a key word or phrase that helps you know which operation to use to answer each question below. State the operation. Why do you think the key word or phrase indicates the operation you chose?**

- **Write an expression you can use to answer the question.**

- **Find the value of your expression.**

a. What is the total amount of trash collected from 2010 to 2013?

b. How many more pounds of recyclables were collected in 2013 than in 2010?

1.1 **Whole Number Operations** (continued)

   c. How many times more recyclables were collected in 2012 than in 2010?

   d. The amount of trash collected in 2014 is estimated to be twice the amount collected in 2011. What is that amount?

2 **ACTIVITY:** Checking Answers

Work with a partner.

   a. Explain how you can use estimation to check the reasonableness of the value of your expression in Activity 1(a).

   b. Explain how you can use addition to check the value of your expression in Activity 1(b).

   c. Explain how you can use estimation to check the reasonableness of the value of your expression in Activity 1(c).

   d. Use mental math to check the value of your expression in Activity 1(d). Describe your strategy.

1.1 **Whole Number Operations** (continued)

3 **ACTIVITY:** Using Estimation

Work with a partner. Use the map. Explain how you found each answer.

a. Which two lakes have a combined area of about 33,000 square miles?

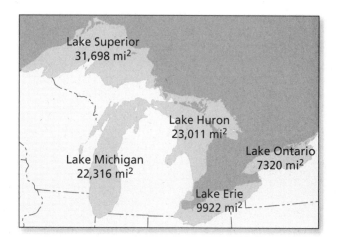

Lake Superior
31,698 mi²

Lake Huron
23,011 mi²

Lake Ontario
7320 mi²

Lake Michigan
22,316 mi²

Lake Erie
9922 mi²

b. Which lake covers an area about three times greater than the area of Lake Erie?

c. Which lake covers an area that is about 16,000 square miles greater than the area of Lake Ontario?

d. Estimate the total area covered by the Great Lakes.

What Is Your Answer?

4. **IN YOUR OWN WORDS** How do you know which operation to choose when solving a real-life problem?

5. In a *magic square*, the sum of the numbers in each row, column, and diagonal is the same and each number from 1 to 9 is used only once. Complete the magic square. Explain how you found the missing numbers.

	9	2
	5	
8		

Name _____ Date _____

Find the value of the expression. Use estimation to check your answer.

1. $5947 + 2001$

2. $\begin{array}{r} 2587 \\ +\ 1654 \\ \hline \end{array}$

3. $5684 + 3118$

4. $1596 - 302$

5. $9564 - 7581$

6. $\begin{array}{r} 7094 \\ -\ 989 \\ \hline \end{array}$

7. $851 \div 37$

8. $\dfrac{612}{68}$

9. $8970 \div 345$

10. $\dfrac{5424}{52}$

11. $8549 \div 198$

12. $74{,}386 \div 874$

13. Your family is traveling 345 miles to an amusement park. You have already traveled 131 miles. How many more miles must you travel to the amusement park?

Name_____ Date _____

1.2 Powers and Exponents
For use with Activity 1.2

Essential Question How can you use repeated factors in real-life situations?

> *As I was going to St. Ives*
> *I met a man with seven wives*
> *Each wife had seven sacks*
> *Each sack had seven cats*
> *Each cat had seven kits*
> *Kits, cats, sacks, wives*
> *How many were going to St. Ives?*   Nursery Rhyme, 1730

1 ACTIVITY: Analyzing a Math Poem

Work with a partner. Here is a "St. Ives" poem written by two students. Answer the question in the poem.

As I was walking into town
I met a ringmaster with five clowns
Each clown had five magicians
Each magician had five bunnies
Each bunny had five fleas
Fleas, bunnies, magicians, clowns
How many were going into town?

Number of clowns:	5	= _____
Number of magicians:	5×5	= _____
Number of bunnies:	$5 \times 5 \times 5$	= _____
Number of fleas:	$5 \times 5 \times 5 \times 5$	= _____

So, the number of fleas, bunnies, magicians, and clowns is _____.

1.2 Powers and Exponents (continued)

2 **ACTIVITY:** Writing Repeated Factors

Work with a partner. Complete the table.

Repeated Factors	Using an Exponent	Value
a. 4×4		
b. 6×6		
c. $10 \times 10 \times 10$		
d. $100 \times 100 \times 100$		
e. $3 \times 3 \times 3 \times 3$		
f. $4 \times 4 \times 4 \times 4 \times 4$		
g. $2 \times 2 \times 2 \times 2 \times 2 \times 2$		

h. In your own words, describe what the two numbers in the expression 3^5 mean.

1.2 **Powers and Exponents** (continued)

3 **ACTIVITY:** Writing and Analyzing a Math Poem

Work with a partner.

a. Write your own "St. Ives" poem.

b. Draw pictures for your poem.

c. Answer the question in your poem.

d. Show how you can use exponents to write your answer.

What Is Your Answer?

4. **IN YOUR OWN WORDS** How can you use repeated factors in real-life situations? Give an example.

5. **STRUCTURE** Use exponents to complete the table. Describe the pattern.

10	100	1000	10,000	100,000	1,000,000
10^1	10^2				

Name _____ Date _____

1.2 Practice
For use after Lesson 1.2

Write the product as a power.

1. $5 \times 5 \times 5$

2. 13×13

3. $8 \cdot 8 \cdot 8 \cdot 8 \cdot 8 \cdot 8$

4. $12 \cdot 12 \cdot 12 \cdot 12 \cdot 12$

5. $10 \cdot 10 \cdot 10 \cdot 10$

6. $17 \times 17 \times 17$

Find the value of the power.

7. 4^4

8. 9^3

9. 24^2

Determine whether the number is a perfect square.

10. 47

11. 16

12. 121

13. You complete 3 centimeters of a necklace in an hour. Each hour after the first, you triple the length of the necklace. Write an expression using exponents for the length of the necklace after 3 hours. Then find the length.

Name_____ Date_____

1.3 Order of Operations
For use with Activity 1.3

Essential Question What is the effect of inserting parentheses into a numerical expression?

1 **ACTIVITY: Comparing Different Orders**

Work with a partner. Find the value of the expression by using different orders of operations. Are your answers the same? (Circle *yes* or *no*.)

a. Add, then multiply.   Multiply, then add.   Same?

$3 + 4 \times 2 =$ _____ $3 + 4 \times 2 =$ _____   Yes   No

b. Add, then subtract.   Subtract, then add.   Same?

$5 + 3 - 1 =$ _____ $5 + 3 - 1 =$ _____   Yes   No

c. Divide, then multiply.   Multiply, then divide.   Same?

$12 \div 3 \bullet 2 =$ _____ $12 \div 3 \bullet 2 =$ _____   Yes   No

d. Divide, then add.   Add, then divide.   Same?

$16 \div 4 + 4 =$ _____ $16 \div 4 + 4 =$ _____   Yes   No

e. Multiply, then subtract.   Subtract, then multiply.   Same?

$8 \times 4 - 2 =$ _____ $8 \times 4 - 2 =$ _____   Yes   No

f. Multiply, then divide.   Divide, then multiply.   Same?

$8 \bullet 4 \div 2 =$ _____ $8 \bullet 4 \div 2 =$ _____   Yes   No

1.3 Order of Operations (continued)

g. Subtract, then add.   Add, then subtract.   Same?

$13 - 4 + 6 =$ _____ $13 - 4 + 6 =$ _____   Yes   No

h. Multiply, then add.   Add, then multiply.   Same?

$1 \times 2 + 3 =$ _____ $1 \times 2 + 3 =$ _____   Yes   No

2 ACTIVITY: Using Parentheses

Work with a partner. Use all the symbols and numbers to write an expression that has the given value.

Symbols and Numbers	Value	Expression
a. $(\ ), +, \div, 3, 4, 5$	3	_____
b. $(\ ), -, \times, 2, 5, 8$	11	_____
c. $(\ ), \times, \div, 4, 4, 16$	16	_____
d. $(\ ), -, \div, 3, 8, 11$	1	_____
e. $(\ ), +, \times, 2, 5, 10$	70	_____

1.3 Order of Operations (continued)

3 **ACTIVITY:** Reviewing Fractions and Decimals

Work with a partner. Evaluate the expression.

a. $\dfrac{3}{4} - \left(\dfrac{1}{4} + \dfrac{1}{2} \right)$

b. $\left(\dfrac{5}{6} - \dfrac{1}{6} \right) - \dfrac{1}{12}$

c. $7.4 - (3.5 - 3.1)$

d. $10.4 - (8.6 + 0.9)$

e. $(\$7.23 + \$2.32) - \$5.40$

f. $\$124.60 - (\$72.41 + \$5.67)$

What Is Your Answer?

4. In an expression with two or more operations, why is it necessary to agree on an order of operations? Give examples to support your explanation.

5. IN YOUR OWN WORDS What is the effect of inserting parentheses into a numerical expression?

1.3 Practice
For use after Lesson 1.3

PEMDAS + 6 At
─────
6

Evaluate the expression.

1. $9 - \boxed{6 \div 3}$

$9 - 2$

$\boxed{7}$

2. $36 - \boxed{7(2)}$

$36 - 14$

$\boxed{22}$

3. $\boxed{(5 + 1)} \div 2$

$6 \div 2$

$\boxed{3}$

4. $8 + \boxed{(10 - 4)} - 3^2$

$8 + 6 - \boxed{3^2}$

$\boxed{8 + 6} - 9$

$14 - 9$

$\boxed{5}$

5. $\boxed{(3 + 5)}^2 \div 4 + 19$

$8^2 \div 4 + 19$

$\boxed{64} \div 4 + 19$

$16 + 19$

$\boxed{35}$

6. $12\boxed{(3 + 3)} \div 18$

$12\boxed{(6)} \div 18$

$72 \div 18$

$\boxed{4}$

7. $\dfrac{(2^2 + 1)}{5}$

1

8. $\dfrac{2(3 + 1)}{8}$

1

9. $\dfrac{10^2 \div 4}{3 + 2}$

5

10. You and three friends go to a restaurant for dinner. You share three appetizers that cost $6 each. You also share two desserts that cost $3 each. You split the total bill evenly. How much does each person pay? $6 \$

1.4 Prime Factorization
For use with Activity 1.4

Essential Question Without dividing, how can you tell when a number is divisible by another number?

1 **ACTIVITY:** Finding Divisibility Tests for 2, 3, 5, and 10

Work with a partner.

a. Highlight all the numbers that are divisible by 2.

b. Put a box around the numbers that are divisible by 3.

c. Underline the numbers that are divisible by 5.

d. Circle the numbers that are divisible by 10.

e. **STRUCTURE** In parts (a)–(d), what patterns do you notice? Write four rules to determine when a number is divisible by 2, 3, 5, and 10.

if you skip count by the number you need you'll get the answer

all evens ÷ 2 : all things ending in 0,5

sum of digets ÷ 3 all numbers ending
with 0 ÷ 10

1.4 **Prime Factorization** (continued)

2 **ACTIVITY:** Finding Divisibility Rules for 6 and 9

Work with a partner.

   a. List ten numbers that are divisible by 6. Write a rule to determine when a number is divisible by 6. Use a calculator to check your rule with large numbers.

   b. List ten numbers that are divisible by 9. Write a rule to determine when a number is divisible by 9. Use a calculator to check your rule with large numbers.

3 **ACTIVITY:** Rewriting a Number Using 2s, 3s, and 5s

Work with three other students. Use the following rules and only the prime factors 2, 3, and 5 to write each number on the next page as a product.

- Your group should have four sets of cards: a set with all 2s, a set with all 3s, a set with all 5s, and a set of blank cards. Each person gets one set of cards.*

- Begin by choosing two cards to represent the given number as a product of two factors. The person with the blank cards writes any factors that are not 2, 3, or 5.

- Use the cards again to represent any number written on a blank card as a product of two factors. Continue until you have represented each handwritten card as a product of two prime factors.

- You may use only one blank card for each step.

*Cut-outs are available in the back of the Record and Practice Journal.

1.4 **Prime Factorization** (continued)

   a. 108   **b.** 80

   c. 162   **d.** 300

   e. Compare your results with those of other groups. Are your steps the same for each number? Is your final answer the same for each number?

What Is Your Answer?

4. IN YOUR OWN WORDS Without dividing, how can you tell when a number is divisible by another number? Give examples to support your explanation.

5. Explain how you can use your divisibility rules from Activities 1 and 2 to help with Activity 3.

1.4 Practice
For use after Lesson 1.4

List the factor pairs of the number.

1. 6

2. 7

3. 10

4. 16

5. 35

6. 55

Write the prime factorization of the number.

7. 9

8. 24

9. 40

10. 44

11. 50

12. 65

13. A fitness instructor arranges 30 people into rows. Each row has the same number of people.

  a. Can the instructor arrange the people into rows of 6?

  b. Can the instructor arrange the people into rows of 9?

Name_____ Date_____

1.5 Greatest Common Factor
For use with Activity 1.5

Essential Question How can you find the greatest common factor of two numbers?

A **Venn diagram** uses circles to describe relationships between two or more sets. The Venn diagram shows the names of students enrolled in two activities. Students enrolled in both activities are represented by the overlap of the two circles.

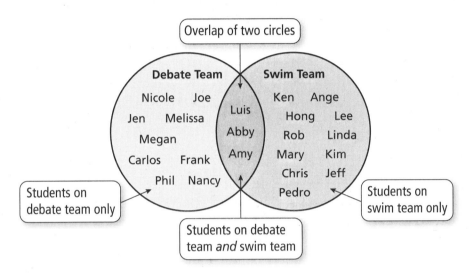

1 ACTIVITY: Identifying Common Factors

Work with a partner. Complete the Venn diagram. Identify the *common factors* of the two numbers.

a. 36 and 48

b. 16 and 56

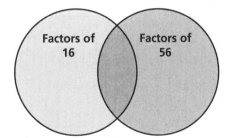

1.5 **Greatest Common Factor** (continued)

c. 30 and 75

d. 54 and 90

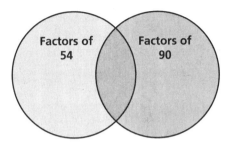

e. Look at the Venn diagrams in parts (a)–(d). Explain how to identify the *greatest common factor* of each pair of numbers. Then circle it in each diagram.

2 **ACTIVITY:** Interpreting a Venn Diagram of Prime Factors

Work with a partner. The Venn diagram represents the prime factorization of two numbers. Identify the two numbers. Explain your reasoning.

a.

b.

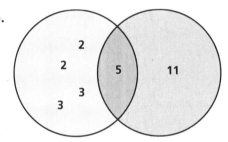

3 **ACTIVITY:** Identifying Common Prime Factors

Work with a partner.

a. Write the prime factorizations of 36 and 48. Use the results to complete the Venn diagram.

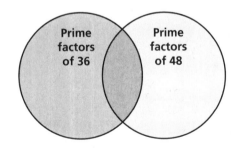

1.5 **Greatest Common Factor** (continued)

b. Repeat part (a) for the remaining number pairs in Activity 1.

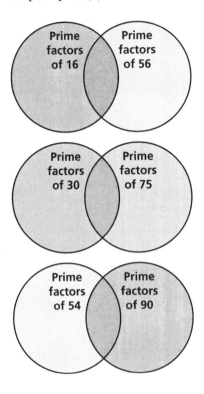

c. **STRUCTURE** Compare the numbers in the overlap of the Venn diagrams to your results in Activity 1. What conjecture can you make about the relationship between these numbers and your results in Activity 1?

What Is Your Answer?

4. **IN YOUR OWN WORDS** How can you find the greatest common factor of two numbers? Give examples to support your explanation.

5. Can you think of another way to find the greatest common factor of two or more numbers? Explain.

1.5 Practice
For use after Lesson 1.5

Find the GCF of the numbers using lists of factors.

1. 9, 15   3

2. 11, 19   1

3. 8, 28   4   6

4. 60, 70   10

5. 40, 56   8

6. 35, 72   1

Find the GCF of the numbers using prime factorizations.

7. 4, 10

8. 5, 11

9. 6, 8

10. 14, 42

11. 45, 63

12. 60, 90

13. You are making identical gift bags using 24 candles and 36 bottles of lotion.
What is the greatest number of gift bags you can make with no items left over?

Name_____ Date _____

Essential Question How can you find the least common multiple of two numbers?

1 ACTIVITY: Identifying Common Multiples

Work with a partner. Using the first several multiples of each number, complete the Venn diagram. Identify any *common multiples* of the two numbers.

a. 8 and 12

b. 4 and 14

c. 10 and 15

d. 20 and 35

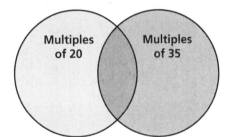

e. Look at the Venn diagrams in parts (a)–(d). Explain how to identify the *least common multiple* of each pair of numbers. Then circle it in each diagram.

Name _____ Date _____

2 **ACTIVITY:** Interpreting a Venn Diagram of Prime Factors

Work with a partner

a. Write the prime factorizations of 8 and 12.
Use the results to complete the Venn diagram.

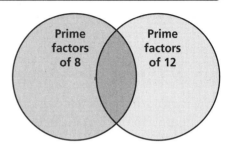

b. Repeat part (a) for the remaining number pairs in Activity 1.

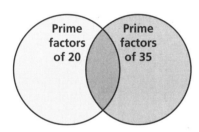

c. **STRUCTURE** Compare the numbers from each section of the Venn diagrams to your results in Activity 1. What conjecture can you make about the relationship between these numbers and your results in Activity 1?

What Is Your Answer?

3. **IN YOUR OWN WORDS** How can you find the least common multiple of two numbers? Give examples to support your explanation.

1.6 **Least Common Multiple** (continued)

4. The Venn diagram shows the prime factors of two numbers. Use the diagram to do the following tasks.

   a. Identify the two numbers.

   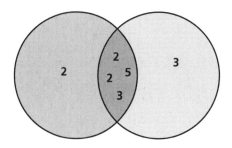

   b. Find the greatest common factor.

   c. Find the least common multiple.

5. A student writes the prime factorizations of 8 and 12 in a table as shown. She claims she can use the table to find the greatest common factor and the least common multiple of 8 and 12. How is this possible?

   $$8 = \begin{array}{|c|c|c|c|} \hline 2 & 2 & 2 & \\ \hline \end{array}$$
   $$12 = \begin{array}{|c|c|c|c|} \hline 2 & 2 & & 3 \\ \hline \end{array}$$

6. Can you think of another way to find the least common multiple of two or more numbers? Explain.

Name _____ Date _____

1.6 Practice
For use after Lesson 1.6

$+\dfrac{13}{13}$ A+

Find the LCM of the numbers using lists of multiples.

1. 3, 8   24

2. 8, 14   56

3. 7, 21   21

4. 5, 11   55

5. 8, 20   40

6. 14, 20   140

Find the LCM of the numbers using prime factorizations.

7. 12, 36   36

8. 5, 12   60

9. 3, 17   51

10. 10, 12   60

11. 20, 30   60

12. 32, 40   160

13. A music store gives every 20th customer a $5 gift card. Every 50th customer gets a $10 gift card. Which customer will be the first to receive both types of gift cards?

the 100th customer

Extension 1.6 Practice
For use after Extension 1.6

Use the LCD to rewrite the fractions with the same denominator.

1. $\dfrac{5}{6}, \dfrac{3}{10}$

2. $\dfrac{5}{9}, \dfrac{11}{12}$

Complete the statement using <, >, or =.

3. $\dfrac{3}{10} \underline{\quad} \dfrac{4}{15}$

4. $\dfrac{1}{2} \underline{\quad} \dfrac{5}{6}$

5. $\dfrac{1}{3} \underline{\quad} \dfrac{4}{12}$

6. $\dfrac{1}{9} \underline{\quad} \dfrac{2}{3}$

Add. Write the answer in simplest form.

7. $\dfrac{2}{3} + \dfrac{5}{12}$

8. $\dfrac{1}{2} + \dfrac{3}{8}$

9. $2\dfrac{5}{7} + 1\dfrac{1}{4}$

10. $3\dfrac{4}{5} + 2\dfrac{1}{2}$

Extension 1.6 Practice (continued)

Subtract. Write the answer in simplest form.

11. $\dfrac{3}{4} - \dfrac{1}{2}$

12. $\dfrac{4}{5} - \dfrac{5}{12}$

13. $4\dfrac{6}{7} - \dfrac{1}{4}$

14. $2\dfrac{7}{9} - 2\dfrac{1}{3}$

15. A recipe calls for $\dfrac{3}{4}$ cup of vegetable broth. You have $\dfrac{2}{3}$ cup of vegetable broth. How much additional broth is needed for the recipe?

16. You have $2\dfrac{3}{4}$ pounds of taffy. You eat $\dfrac{1}{3}$ pound of taffy. How many pounds of taffy do you have left?

Chapter 2 Fair Game Review

Estimate the product or quotient.

1. 91×17

2. 57×29

3. $83 \div 18$

4. $204 \div 9$

5. $152 \div 31$

6. 13×78

7. 32×51

8. $651 \div 49$

9. There are 546 people attending a charity event. You are baking cookies to give away. Each batch makes 48 cookies. Estimate the number of batches you need to make so that each person gets one cookie.

Chapter 2 **Fair Game Review** (continued)

Find the product or quotient.

10. 351
    $\times$ 15

11. 187
    $\times$ 27

12. $9\overline{)333}$

13. $3\overline{)474}$

14. A bleacher row can seat 14 people. The bleachers are filled to capacity with 1330 people at a soccer game. How many rows of bleachers does the soccer field have?

Name_____ Date _____

Essential Question What does it mean to multiply fractions?

1 **ACTIVITY:** Multiplying Fractions

Work with a partner. A bottle of water is $\frac{1}{2}$ full. You drink $\frac{2}{3}$ of the water.

How much of the bottle of water do you drink?

THINK ABOUT THE QUESTION: To help you think about this question, rewrite the question.

Words: What is $\frac{2}{3}$ of $\frac{1}{2}$? **Numbers:** $\frac{2}{3} \times \frac{1}{2} = ?$

Here is one way to get the answer.

- **Draw** a segment to represent a length of $\frac{1}{2}$.

- **Show** how to divide $\frac{1}{2}$ into three equal parts.

- **Rewrite** $\frac{1}{2}$ as a fraction whose numerator is divisible by 3.

- Each part is $\frac{1}{6}$ of the bottle of water, and you drank two of them. Written

  as multiplication, you have $\frac{2}{3} \times \frac{1}{2} =$ _____.

You drank _____ of the bottle of water.

Name _____ Date _____

2 **ACTIVITY:** Multiplying Fractions

A park has a playground that is $\frac{3}{4}$ of its width and $\frac{4}{5}$ of its

length. What fraction of the park is covered by the playground?

Fold a piece of paper horizontally into fourths and shade three

of the fourths to represent $\frac{3}{4}$.

Fold the paper vertically into fifths and shade $\frac{4}{5}$ of the paper

another color.

Count the total number of squares. This number is the denominator.
The numerator is the number of squares shaded with both colors.

$\frac{3}{4} \times \frac{4}{5} =$ _____ = _____. So, _____ of the park is covered by the playground.

Inductive Reasoning

Work with a partner. Complete the table by using a model or folding paper.

	Exercise	Verbal Expression	Answer
1	3. $\frac{2}{3} \times \frac{1}{2}$		
2	4. $\frac{3}{4} \times \frac{4}{5}$		
	5. $\frac{2}{3} \times \frac{5}{6}$		
	6. $\frac{1}{6} \times \frac{1}{4}$		
	7. $\frac{2}{5} \times \frac{1}{2}$		
	8. $\frac{5}{8} \times \frac{4}{5}$		

2.1 **Multiplying Fractions** (continued)

What Is Your Answer?

9. **IN YOUR OWN WORDS** What does it mean to multiply fractions?

10. **STRUCTURE** Write a general rule for multiplying fractions.

2.1 Practice
For use after Lesson 2.1

Multiply. Write the answer in simplest form.

1. $\dfrac{1}{6} \times \dfrac{5}{8}$

2. $\dfrac{7}{9} \times 3$

3. $\dfrac{8}{9} \times \dfrac{3}{5}$

4. $\dfrac{7}{8} \times 2\dfrac{1}{3}$

5. $7 \times 3\dfrac{9}{14}$

6. $5\dfrac{5}{9} \times 2\dfrac{7}{10}$

7. You reserve $\dfrac{2}{5}$ of the seats on a tour bus. You are able to fill $\dfrac{5}{8}$ of the seats you reserve. What fraction of the seats on the bus are you able to fill?

8. A triangle has a base of $5\dfrac{2}{3}$ inches and a height of 3 inches. What is the area of the triangle?

2.2 Dividing Fractions
For use with Activity 2.2

Essential Question How can you divide by a fraction?

---

1 ACTIVITY: Dividing by a Fraction

Work with a partner. Write the division problem and solve it using a model.

  a. How many two-thirds are in three?

The division problem is _____.

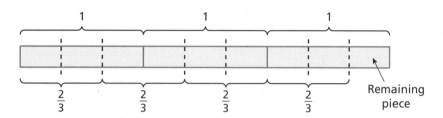

How many groups of $\dfrac{2}{3}$ are in 3? _____

The remaining piece represents _____ of $\dfrac{2}{3}$.

So, there are _____ groups of $\dfrac{2}{3}$ in 3.

So, _____ ÷ _____ = _____.

  b. How many halves are in five halves?

  c. How many four-fifths are in eight?

2.2 **Dividing Fractions** (continued)

    d. How many one-thirds are in seven halves?

    e. How many three-fourths are in five halves?

2 **ACTIVITY:** Using Tables to Recognize a Pattern

Work with a partner.

    a. Complete each table.

Division Table

$8 \div 16$	
$8 \div 8$	
$8 \div 4$	
$8 \div 2$	
$8 \div 1$	
$8 \div \dfrac{1}{2}$	
$8 \div \dfrac{1}{4}$	
$8 \div \dfrac{1}{8}$	

Multiplication Table

$8 \times \dfrac{1}{16}$	
$8 \times \dfrac{1}{8}$	
$8 \times \dfrac{1}{4}$	
$8 \times \dfrac{1}{2}$	
8×1	
8×2	
8×4	
8×8	

2.2 **Dividing Fractions** (continued)

b. Describe the relationship between the numbers in the right column of the division table and the numbers in the right column of the multiplication table.

c. Describe the relationship between the shaded numbers in the division table and the shaded numbers in the multiplication table.

d. **STRUCTURE** Make a conjecture about how you can use multiplication to divide by a fraction.

e. Test your conjecture using the problems in Activity 1.

What Is Your Answer?

3. **IN YOUR OWN WORDS** How can you divide by a fraction? Give an example.

4. How many halves are in a fourth? Explain how you found your answer.

Name _____ Date _____

Complete the statement.

1. $\dfrac{3}{8} \times$ ____ $= 1$

2. $7 \times$ ____ $= 1$

3. $3 \div$ ____ $= 36$

4. $\dfrac{4}{9} \div$ ____ $= 12$

Evaluate the expression.

5. $\dfrac{1}{3} \div \dfrac{1}{6}$

6. $\dfrac{3}{8} \div \dfrac{5}{8}$

7. $6 \div \dfrac{2}{5}$

8. $\dfrac{4}{9} \div \dfrac{2}{3} \div \dfrac{5}{6}$

9. $\dfrac{1}{3} + \dfrac{4}{7} \div \dfrac{3}{10}$

10. $\dfrac{7}{8} \bullet \dfrac{4}{5} \div \dfrac{7}{20}$

11. In a jewelry store, rings make up $\dfrac{5}{9}$ of the inventory. Earrings make up $\dfrac{4}{15}$ of the inventory. How many times greater is the ring inventory than the earring inventory?

2.3 Dividing Mixed Numbers
For use with Activity 2.3

Essential Question How can you model division by a mixed number?

---

1 ACTIVITY: Writing a Story

Work with a partner. Think of a story that uses division by a mixed number.

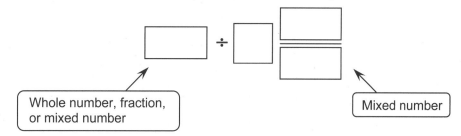

Whole number, fraction, or mixed number

Mixed number

a. Write your story. Then draw pictures for your story.

b. Solve the division problem and use the answer in your story. Include a diagram of the division problem.

There are many possible stories. Here is one that uses $6 \div 1\frac{1}{2}$.

Joe goes on a camping trip with his aunt, his uncle, and three cousins. They leave at 5:00 P.M. and drive 2 hours to the campground.

Joe helps his uncle put up three tents. His aunt cooks hamburgers on a grill that is over a fire.

In the morning, Joe tells his aunt that he is making pancakes. He decides to triple the recipe so there will be plenty of pancakes for everyone. A single recipe uses 2 cups of water, so he needs a total of 6 cups.

Pancake Mix
Recipe:
2 cups water
2 cups pancake mix
1/4 cup oil
1 egg
1/4 teaspoon salt

2.3 **Dividing Mixed Numbers** (continued)

Joe's aunt has a 1-cup measuring cup and a ½-cup measuring cup. The water faucet is about 50 yards from the campsite. Joe tells his cousins that he can get 6 cups of water in only 4 trips.

When his cousins ask him how he knows that, he uses a stick to draw a diagram in the dirt. Joe says, "This diagram shows that there are four 1½s in 6." In other words,

$$6 \div 1\frac{1}{2} = 4.$$

2 **ACTIVITY:** Dividing Mixed Numbers

Work with a partner. Write the division problem and solve it using a model.

a. How many three-fourths are in four and one-half?

b. How many five-sixths are in three and one-third?

2.3 Dividing Mixed Numbers (continued)

   c. How many three-eighths are in three and three-fourths?

   d. How many one and one-halves are in six?

   e. How many one and one-fifths are in five?

   f. How many one and one-fourths are in four and one-half?

   g. How many two and one-thirds are in five and five-sixths?

What Is Your Answer?

   3. IN YOUR OWN WORDS How can you model division by a mixed number?

   4. Can you think of another method you can use to obtain your answers in Activity 2?

2.3 Practice
For use after Lesson 2.3

Divide. Write the answer in simplest form.

1. $4\dfrac{1}{6} \div 5$

2. $\dfrac{5}{8} \div 5\dfrac{3}{4}$

3. $8\dfrac{1}{6} \div 2\dfrac{1}{24}$

4. $2\dfrac{3}{10} \div 3\dfrac{3}{5}$

5. $6\dfrac{6}{7} \div 3\dfrac{3}{5}$

6. $3\dfrac{3}{5} \div 6\dfrac{6}{7}$

Evaluate the expression.

7. $4\dfrac{7}{12} \div \dfrac{3}{4} \times \dfrac{3}{11}$

8. $9 \div 8\dfrac{1}{10} - \dfrac{5}{9}$

9. $5\dfrac{7}{8} \times \left(2\dfrac{4}{5} \div 7\right)$

10. At a road race, you have $60\dfrac{3}{4}$ feet available for a water station. Your tables are $6\dfrac{3}{4}$ feet long. How many tables can you line up for the water station?

2.4 Adding and Subtracting Decimals
For use with Activity 2.4

Essential Question How can you add and subtract decimals?

Base ten blocks can be used to model numbers.*

1 one   1 tenth   1 hundredth

---

1 **ACTIVITY: Modeling a Sum**

Work with a partner. Use base ten blocks to find the sum.

   a. 1.23 + 0.87

Which base ten blocks do you need to model the numbers in the sum?
How many of each do you need? Draw a sketch of your model.

How many of each base ten block do you have when you combine
the blocks?

_____ ones   _____ tenths   _____ hundredths

How many of each base ten block do you have when you trade the blocks?

_____ ones   _____ tenths   _____ hundredths

   So, 1.23 + 0.87 = _____.

 b. 1.25 + 1.35   **c.** 2.14 + 0.92   **d.** 0.73 + 0.86

*Cut-outs are available in the back of the Record and Practice Journal.

2.4 **Adding and Subtracting Decimals** (continued)

2 **ACTIVITY:** Modeling a Difference

Work with a partner. Use base ten blocks to find the difference.

a. 2.43 − 0.73

Which number is shown by the model?

Circle the portion of the model that represents 0.73.

So, 2.43 − 0.73 = _____.

b. 1.86 − 1.26   **c.** 3.72 − 0.5   **d.** 1.58 − 0.09

3 **ACTIVITY:** Making a Conjecture

Work with a partner.

a. Find each sum or difference.

| 123 + 87 | 125 + 135 | 214 + 92 | 73 + 86 |
| 243 − 73 | 186 − 126 | 372 − 50 | 158 − 9 |

b. How are the numerical expressions in part (a) related to the numerical expressions in Activities 1 and 2? How are the sums and differences related?

c. **STRUCTURE** There is a relationship between adding and subtracting decimals and adding and subtracting whole numbers. What conjecture can you make about this relationship?

2.4 **Adding and Subtracting Decimals** (continued)

4 **ACTIVITY:** Using a Place Value Chart

Work with a partner. Use the place value chart to find the sum or difference.

	Place Value Chart	

millions | hundred thousands | ten thousands | thousands | hundreds | tens | ones | and | tenths | hundredths | thousandths | ten-thousandths | hundred-thousandths | millionths

a. $16.05 + 2.94$

b. $7.421 + 92.55$

c. $38.72 - 8.61$

d. $64.968 - 51.167$

What Is Your Answer?

5. MODELING Describe two real-life examples of when you would need to add and subtract decimals.

6. IN YOUR OWN WORDS How can you add and subtract decimals?

Name _____ Date _____

Add.

1. $3.02 + 1.67$

2. $1.4 + 8.68$

3. $11.514 + 4.29$

4. $15.71 + 12.643$

5. $9.562 + 21.764$

6. $15.602 + 2.47$

Subtract.

7. $2.64 - 1.52$

8. $4.023 - 3.146$

9. $7.87 - 5.152$

10. $16.045 - 12.63$

11. $17.1 - 11.457$

12. $5.18 - 2.487$

13. You buy a movie for $19.99 and a set of earphones for $12.49. How much is the bill before taxes?

Name_____ Date_____

2.5 Multiplying Decimals
For use with Activity 2.5

Essential Question   How can you multiply decimals?

1 **ACTIVITY:** Multiplying Decimals Using a Rectangle

Work with a partner. Use a rectangle to find the product.

  a. $2.7 \cdot 1.3$

  Arrange base ten blocks to form a rectangle of length 2.7 units and width 1.3 units. Sketch your model.

  The area of the rectangle represents the product.

  Find the total area represented by each grouping of base ten blocks.

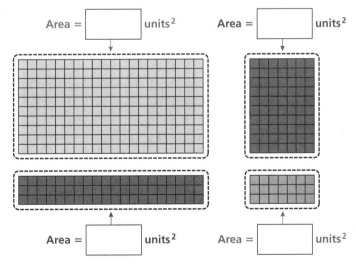

Area = [____] units²   Area = [____] units²

Area = [____] units²   Area = [____] units²

  The area of the rectangle is

  ____ + ____ + ____ + ____ = ____ units².

  So, $2.7 \cdot 1.3 =$ _____.

2.5 **Multiplying Decimals** (continued)

    b. $1.8 \cdot 1.1$ **c.** $4.6 \cdot 1.2$ **d.** $3.2 \cdot 2.4$

2   **ACTIVITY:** Multiplying Decimals Using an Area Model

Work with a partner. Use an area model to find the product. Explain your reasoning.

    a. $0.8 \cdot 0.5$

        Use the 10-by-10 square grid.

        Shade 8 rows of the grid to represent 0.8.

        Shade 5 columns of the grid to represent 0.5. Use a different color.

        Because _____ hundredths are shaded with

        both colors, the product is $\dfrac{}{100}$ = _____.

        So, $0.8 \cdot 0.5 =$ _____.

    b. $0.3 \cdot 0.5$ **c.** $0.7 \cdot 0.6$ **d.** $0.2 \cdot 0.9$

2.5 **Multiplying Decimals** (continued)

3 **ACTIVITY:** Making a Conjecture

Work with a partner.

a. Find each product.

27 • 13   18 • 11   46 • 12   32 • 24

8 • 5   3 • 5   7 • 6   2 • 9

b. How are the numerical expressions in part (a) related to the numerical expressions in Activities 1 and 2? How are the products related?

c. **STRUCTURE** What conjecture can you make about the relationship between multiplying decimals and multiplying whole numbers?

What Is Your Answer?

4. **IN YOUR OWN WORDS** How can you multiply decimals?

Name _____ Date _____

2.5 Practice
For use after Lesson 2.5

Multiply. Use estimation to check your answer.

1.　0.5
　　× 4

2.　3.8
　　× 6

3.　2.1
　　× 11

4.　0.8
　　× 0.6

5.　0.003
　　× 0.09

6.　8.91
　　× 1.26

7. You earn $7.80 an hour working as a dog sitter. You work 12.5 hours during the weekend. How much money do you make?

8. You use a microscope to look at bacteria that is 0.0034 millimeter long. The microscope magnifies the bacteria 430 times. How long does the bacteria appear to be when you look at it through the microscope?

Name_____ Date_____

2.6 Dividing Decimals
For use with Activity 2.6

Essential Question How can you use base ten blocks to model decimal division?

1 ACTIVITY: Dividing Decimals

Work with a partner. Use base ten blocks to model the division.

  a. $2.4 \div 0.6$

    Begin by modeling 2.4 with base ten blocks. Sketch your model.

    How many of each base ten block did you use?

    _____ ones   _____ tenths   _____ hundredths

    Next, think of the division problem $2.4 \div 0.6$ as the question,

       "How can you divide 2.4 into groups of 0.6 ?"

    Rearrange the model for 2.4 into groups of 0.6. Sketch your model.

      There are _____ groups of 0.6. So, $2.4 \div 0.6 = $ _____.

2.6 **Dividing Decimals** (continued)

    b. $1.8 \div 2$ **c.** $3.9 \div 3$

    d. $2.8 \div 0.7$ **e.** $3.2 \div 0.4$

    f. Write and solve the division problem represented by the model.

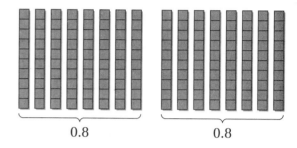

                               0.8   0.8

2 **ACTIVITY:** Dividing Decimals

Work with a partner. Use base ten blocks to model the division.

    a. $0.3 \div 0.06$

      Model 0.3.   Replace tenths with hundredths.   How many 0.06s are in 0.3? Divide hundredths into groups of 0.06.

      There are _____ groups of 0.06. So, $0.3 \div 0.06 =$ _____.

    b. $0.2 \div 0.04$ **c.** $0.6 \div 0.01$

    d. $0.16 \div 0.08$ **e.** $0.28 \div 0.07$

2.6 Dividing Decimals (continued)

What Is Your Answer?

3. **IN YOUR OWN WORDS** How can you use base ten blocks to model decimal division? Use examples from Activity 1 and Activity 2 as part of your answer.

4. **WRITING** Newton's poem is about dividing fractions. Write a poem about dividing decimals.

"When you must divide a fraction, do this very simple action:
Flip what you're dividing BY, and then it's easy—multiply!"

5. Think of your own cartoon about dividing decimals. Draw your cartoon.

2.6 Practice
For use after Lesson 2.6

Divide. Check your answer.

1. $3\overline{)18.6}$

2. $6\overline{)46.8}$

3. $4\overline{)7.6}$

4. $24.5 \div 7$

5. $0.096 \div 8$

6. $15.65 \div 5$

7. $3.1\overline{)17.36}$

8. $6.4\overline{)43.52}$

9. $7.05\overline{)8.46}$

10. $9.24 \div 15.4$

11. $7.06 \div 0.353$

12. $0.015 \div 0.003$

13. It costs $859.32 to have a school dance.

    a. How many tickets must be sold to cover the cost?

School Dance

October 28th
Tickets $8

    b. How many tickets must be sold to make a $980.68 profit?

Chapter 3 **Fair Game Review**

Write a sentence interpreting the expression.

1. $2 \times (126 + 2566)$

2. $4 \times (6425 + 25)$

3. $(65 - 23) + 3$

4. $(65{,}000 - 5169) + 58$

5. $(890 \div 2) \div 2$

6. $(65 \times 6) \div 3$

7. Write a real-life problem representing the expression below.

$$3 \times (20 + 6)$$

Chapter 3 **Fair Game Review** (continued)

Simplify the expression.

8. $4 - 8 \div 2$

9. $2^2 \cdot 3 - 3$

10. $16 - 32 \div 2^3$

11. $3(4^2 - 9)$

12. $12 + 16 \div 4 \cdot 2$

13. $24 - 18 \div 3 + 2$

14. $20 + 12 \div 2(7 - 4)$

15. $4(3^3 - 7) \div 10$

16. A group of 4 adults and 5 children is visiting an amusement park. Admission is \$15 per adult and \$9 per child. Find the total cost of admission for the group.

3.1 Algebraic Expressions
For use with Activity 3.1

Essential Question How can you write and evaluate an expression that represents a real-life problem?

1 ACTIVITY: Reading and Re-Reading

a.

> You babysit for 3 hours. You receive $12. What is your hourly wage?

- Underline the important numbers and units you need to solve the problem.

- Read the problem carefully a second time. Circle the key word for the question.

- Write each important number or word, with its units, on a piece of paper. Write $+$, $-$, $\times$, $\div$, and $=$ on five other pieces of paper.

- Arrange the pieces of paper to answer the key word question, "What is your hourly wage?"

- Evaluate the expression that represents the hourly wage.

Your hourly wage is _____.

b. How can you use your hourly wage to find how much you will receive for any number of hours worked?

Name _____ Date _____

2 ACTIVITY: Reading and Re-Reading

Work with a partner. Use the strategy shown in Activity 1 to write an expression for each problem. After you have written the expression, evaluate it using mental math or some other method.

a. You wash cars for 2 hours. You receive $6.
   How much do you earn per hour?

   Expression: _____

   Amount you earn per hour: _____

b. You have $60. You buy a pair of jeans and a shirt.
   The pair of jeans costs $27. You come home with
   $15. How much did you spend on the shirt?

   Expression: _____

   Amount you spend on shirt: _____

c. For lunch, you buy 5 sandwiches that cost $3 each.
   How much do you spend?

   Expression: _____

   Amount you spend on sandwiches: _____

3.1 **Algebraic Expressions** (continued)

d. You are running a 4500-foot race. How much farther do you have to go after running 2000 feet?

Expression: _____

Amount left to go: _____

e. A young rattlesnake grows at a rate of about 20 centimeters per year. How much does a young rattlesnake grow in 2 years?

Expression: _____

Amount rattlesnake grows in 2 years: _____

What Is Your Answer?

3. IN YOUR OWN WORDS How can you write and evaluate an expression that represents a real-life problem? Give one example with addition, one with subtraction, one with multiplication, and one with division.

3.1 Practice
For use after Lesson 3.1

Evaluate the expression when $a = 4$, $b = 5$, and $c = 10$.

1. $a + 7$

2. $b - 3$

3. $9c$

4. $25 \div b$

5. $a \bullet c$

6. $b - a$

7. $a + b + c$

8. $\dfrac{c}{b}$

9. $4a - 7$

10. You need $2b$ cups of flour for making b loaves of bread. You have 8 cups of flour. Do you have enough for 5 loaves of bread? Explain.

11. The expression $9a + 6s$ is the cost for a adults and s students to see a musical performance.

    a. Find the total cost for three adults and five students.

    b. Find the total cost for four adults and four students.

Name_____ Date_____

Essential Question How can you write an expression that represents an unknown quantity?

1 ACTIVITY: Ordering Lunch

Work with a partner. You use a $20 bill to buy lunch at a café. You order a sandwich from the menu board.

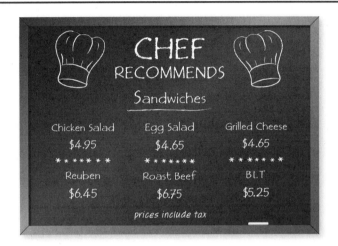

CHEF RECOMMENDS
Sandwiches

Chicken Salad	Egg Salad	Grilled Cheese
$4.95	$4.65	$4.65
Reuben	Roast Beef	BLT
$6.45	$6.75	$5.25

prices include tax

a. Complete the table. In the last column, write a numerical expression for the amount of change received.

Sandwich	Price (dollars)	Change Received (dollars)
Reuben		
BLT		
Egg salad		
Roast beef		

b. REPEATED REASONING Write an expression for the amount of change you receive when you order any sandwich from the menu board.

c. Compare the expression you wrote in part (b) with the expression in the last column of the table in part (a).

3.2 **Writing Expressions** (continued)

d. The café offers several side dishes, each at the same price. You order a chicken salad sandwich and two side dishes. Write an expression for the total amount of money you spend. Explain how you wrote your expression.

e. The expression $20 - 4.65s$ represents the amount of change one customer receives after ordering from the menu board. Explain what each part of the expression represents. Do you know what the customer ordered? Explain your reasoning.

2 **ACTIVITY:** Words That Imply Addition or Subtraction

Work with a partner.

a. Complete the table.

Variable	Phrase	Expression
n	4 more than a number	
m	the difference of a number and 3	
x	the sum of a number and 8	
p	10 less than a number	
n	7 units farther away	
t	8 minutes sooner	
w	12 minutes later	
y	a number increased by 9	

3.2 **Writing Expressions** (continued)

b. Here is a word problem that uses one of the expressions in the table.

You arrive at the café 8 minutes sooner than your friend. Your friend arrives at 6:42 P.M. When did you arrive?

Which expression from the table on the previous page can you use to solve the problem?

c. Write a problem that uses a different expression from the table.

3 **ACTIVITY:** Words That Imply Multiplication or Division

Work with a partner. Match each phrase with an expression.

the product of a number and 3 $n \div 3$

the quotient of 3 and a number $4p$

4 times a number $n \bullet 3$

a number divided by 3 $2m$

twice a number $3 \div n$

What Is Your Answer?

4. IN YOUR OWN WORDS How can you write an expression that represents an unknown quantity? Give examples to support your explanation.

3.2 Practice
For use after Lesson 3.2

Write the phrase as an expression.

1. the total of 8 and 13

2. 42 divided by 7

Give two ways to write the expression as a phrase.

3. $6 + p$

4. $9m$

Write the phrase as an expression. Then evaluate when $x = 3$ and $y = 15$.

5. 7 more than the quotient of a number y and 5

6. twice the sum of a number x and 8

7. You earn $7 for every hour that you babysit.

  a. Complete the table.

Hours	1	2	3	4	5	6	7	8
Earnings								

  b. Write an expression for the amount you earn after h hours.

Name_____ Date_____

3.3 Properties of Addition and Multiplication
For use with Activity 3.3

Essential Question Does the order in which you perform an operation matter?

1 ACTIVITY: Does Order Matter?

Work with a partner. Place each statement in the correct oval.

a. Fasten 5 shirt buttons.

b. Put on a shirt and tie.

c. Fill and seal an envelope.

d. Floss your teeth.

e. Put on your shoes.

f. Chew and swallow.

Order Matters

Order Doesn't Matter

Write three math problems using the four operations where order matters and three where order doesn't matter.

3.3 **Properties of Addition and Multiplication** (continued)

2 **ACTIVITY:** Commutative Properties

Work with a partner.

   a. Circle the statements that are true.

$$3 + 5 \overset{?}{=} 5 + 3 \qquad\qquad 3 - 5 \overset{?}{=} 5 - 3$$

$$9 \times 3 \overset{?}{=} 3 \times 9 \qquad\qquad 9 \div 3 \overset{?}{=} 3 \div 9$$

   b. The true equations show the Commutative Properties of Addition and Multiplication. Why do you think they are called *commutative*?

3 **ACTIVITY:** Associative Properties

Work with a partner.

   a. Circle the statements that are true.

$$8 + (3 + 1) \overset{?}{=} (8 + 3) + 1 \qquad\qquad 8 - (3 - 1) \overset{?}{=} (8 - 3) - 1$$

$$12 \times (6 \times 2) \overset{?}{=} (12 \times 6) \times 2 \qquad\qquad 12 \div (6 \div 2) \overset{?}{=} (12 \div 6) \div 2$$

   b. The true equations show the Associative Properties of Addition and Multiplication. Why do you think they are called *associative*?

What Is Your Answer?

   4. IN YOUR OWN WORDS Does the order in which you perform an operation matter? Give examples to support your explanation.

3.3 **Properties of Addition and Multiplication** (continued)

5. **MENTAL MATH** Explain how you can add the sum in your head.

$11 + 7 + 12 + 13 + 8 + 9$

6. **SECRET CODE** The creatures on a distant planet use the symbols ■,◆,★, and ● for the four operations.

a. Use the codes to decide which symbol represents addition and which symbol represents multiplication. Explain your reasoning.

$$3 ● 4 = 4 ● 3$$

$$3 ★ 4 = 4 ★ 3$$

$$2 ● (5 ● 3) = (2 ● 5) ● 3$$

$$2 ★ (5 ★ 3) = (2 ★ 5) ★ 3$$

$$0 ● 4 = 0$$

$$0 ★ 4 = 4$$

b. Make up your own symbols for addition and multiplication. Write codes using your symbols. Trade codes with a classmate. Decide which symbol represents addition and which symbol represents multiplication.

Name _____ Date _____

Tell which property illustrates the statement.

1. $x \cdot 1 = x$

2. $4.8 + k = k + 4.8$

Simplify the expression. Explain each step.

3. $8 + (7 + x)$

4. $10(11a)$

Complete the statement using the specified property.

	Property	Statement
5.	Addition Property of Zero	$(b + 0) + 6 =$
6.	Commutative Property of Multiplication	$3 \cdot (n \cdot 5) =$

7. You earn 10 points for every coin you collect in a video game. Then you find a star that triples your score.

   a. Write an expression for the number of points you earn from the coins.

   b. Write and simplify an expression for the total number of points you earn.

3.4 The Distributive Property
For use with Activity 3.4

Essential Question How do you use mental math to multiply two numbers?

> **1 ACTIVITY:** Modeling a Property

Work with a partner.

a. **MODELING** Draw two rectangles of the same width but with different lengths. Label the dimensions.

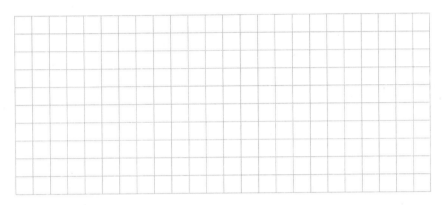

b. Write an expression for the total area of the rectangles.

$$(\underline{\hspace{1cm}} \times \underline{\hspace{1cm}}) + (\underline{\hspace{1cm}} \times \underline{\hspace{1cm}})$$

c. Rearrange the rectangles by aligning the shortest sides to form one rectangle. Label the dimensions. Write an expression for the area.

$$\underline{\hspace{1cm}} \times (\underline{\hspace{1cm}} + \underline{\hspace{1cm}})$$

d. Can the expressions from parts (b) and (c) be set equal to each other? Explain.

e. **REPEATED REASONING** Repeat this activity using different rectangles. Explain how this illustrates the Distributive Property. Write a rule for the Distributive Property.

3.4 The Distributive Property (continued)

2 ACTIVITY: Using Mental Math

Work with a partner. Use mental math to find the product.

   a. 23×6

   b. 33×7 **c.** 47×9

   d. 28×5 **e.** 17×4

3 ACTIVITY: Using Mental Math

Work with a partner. Use the Distributive Property and mental math to find the product.

   a. 6×23

3.4 **The Distributive Property** (continued)

b. 5×17

c. 8×26

d. 20×19

e. 40×29

f. 25×39

g. 15×47

What Is Your Answer?

4. Compare the methods in Activities 2 and 3.

5. IN YOUR OWN WORDS How do you use mental math to multiply two numbers? Give examples to support your explanation.

3.4 Practice
For use after Lesson 3.4

Use the Distributive Property and mental math to find the product.

1. 4×31

2. 7×49

3. $16(38)$

Use the Distributive Property to simplify the expression.

4. $8(5 + w)$

5. $11(9 + d)$

6. $15(p - 4 + 2)$

Simplify the expression.

7. $2x - 4 + 3x$

8. $4y - 1 - 3y + 2$

9. $x + 2(x - 4)$

10. A jazz band sells 31 large boxes of fruit and 74 small boxes of fruit for a fundraiser.

a. Use the Distributive Property to write and simplify an expression for the profit.

Profit = Price − Cost

b. A large box of fruit costs $9 and a small box of fruit costs $4. What is the jazz band's profit?

Extension 3.4 Practice
For use after Extension 3.4

Factor the expression using the GCF.

1. $2 + 8$

2. $9 - 3$

3. $30 + 25$

4. $35 - 14$

5. $81 - 18$

6. $60 + 100$

7. $28 - 20$

8. $72 + 48$

9. $12x + 18$

10. $4y + 10$

**Extension
3.4** **Practice** (continued)

11. $32y - 48$

12. $15y + 40$

13. $16x + 24$

14. $11x + 33$

15. $13x + 39y$

16. $21x - 42y$

17. The length of a rectangle is 6 inches and its area is $(18x + 24)$ square inches. Write an expression for the width.

Name_____ Date _____

Find the area of the square or rectangle.

1.

5 ft

10 ft

2.

3 cm

3 cm

3.

4 in.

4 in.

4.

7 yd

2 yd

5. Find the area of the patio.

8 ft

10 ft

Chapter 4 **Fair Game Review** (continued)

Plot the ordered pair in a coordinate plane.

6. $(2, 3)$

7. $(6, 5)$

8. $(1, 7)$

9. $(4, 4)$

10. $(5, 2)$

11. $(3, 1)$

Name_____ Date _____

4.1   Areas of Parallelograms
For use with Activity 4.1

Essential Question   How can you derive a formula for the area of a parallelogram?

A polygon is a closed figure in a plane that is made up of three or more line segments that intersect only at their endpoints. Several examples of polygons are parallelograms, triangles, and trapezoids.

The formulas for the areas of polygons can be derived from one area formula, the area of a rectangle. Recall that the area of a rectangle is the product of its length ℓ and its width w. The process you use to derive these other formulas is called *deductive reasoning*.

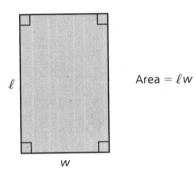

ℓ Area $= \ell w$

w

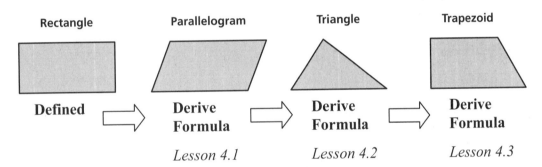

Rectangle	Parallelogram	Triangle	Trapezoid
Defined	**Derive Formula**	**Derive Formula**	**Derive Formula**
	Lesson 4.1	*Lesson 4.2*	*Lesson 4.3*

---

1 ACTIVITY: Deriving the Area Formula of a Parallelogram

Work with a partner.

a. Draw *any* rectangle on a piece of grid paper. An example is shown. Label the length and width. Then find the area of your rectangle.

width = w

right angle — length = ℓ

$A =$ _____

b. Cut your rectangle into two pieces to form a parallelogram. Compare the area of the rectangle with the area of the parallelogram. What do you notice? Use your results to write a formula for the area A of a parallelogram.

4.1 **Areas of Parallelograms** (continued)

2 **ACTIVITY:** Finding the Areas of Parallelograms

Work with a partner.

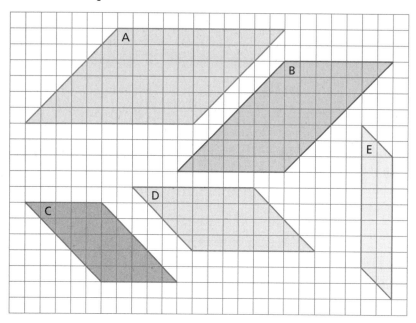

a. Find the area of each parallelogram by cutting it into two pieces to form a rectangle.*

b. Use the formula you wrote in Activity 1 to find the area of each parallelogram. Compare your answers to those in part (a).

c. Count unit squares for each parallelogram to check your results.

*Cut-outs are available in the back of the Record and Practice Journal.

4.1 Areas of Parallelograms (continued)

What Is Your Answer?

3. **IN YOUR OWN WORDS** How can you derive a formula for the area of a parallelogram?

4. **REASONING** The areas of a rectangle and a parallelogram are equal. The length of the rectangle is equal to the base of the parallelogram. What can you say about the width of the rectangle and the height of the parallelogram? Draw a diagram to support your answer.

5. What is the height of the parallelogram shown? How do you know?

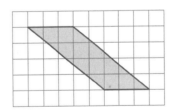

Name _____ Date _____

Find the area of the parallelogram.

1.

12 in.
14 in.

2.

11 cm
5 cm

Find the area of the shaded region.

3.

3 m
2 m
6 m
4 m

4.

4 ft
7 ft
2 ft
10 ft

5. A stained glass window has an area of 900 square inches.

   a. One window design is made of rectangular stained glass pieces that are 5 inches by 3 inches. How many stained glass pieces are used in the window?

   b. Another window design is made of square stained glass pieces that are 6 inches by 6 inches. How many stained glass pieces are used in the window?

Name_____ Date _____

Essential Question How can you derive a formula for the area of a triangle?

1 ACTIVITY: Deriving the Area Formula of a Triangle

Work with a partner.

 a. Draw *any* rectangle on a piece of grid paper. Label the length and width. Then find the area of your rectangle.

 b. Draw a diagonal from one corner of your rectangle to the opposite corner. Cut along the diagonal. Compare the area of the rectangle with the area of the two pieces you cut. What do you notice? Use your results to write a formula for the area *A* of a triangle.

 Area = _____ Formula

2 ACTIVITY: Deriving the Area Formula of a Triangle

Work with a partner.

 a. Fold a piece of grid paper in half. Draw a triangle so that its base lies on one of the horizontal lines of the paper. Do not use a right triangle. Label the height and the base *inside* the triangle.

 b. Estimate the area of your triangle by counting unit squares.

 Area ≈ _____ Estimate

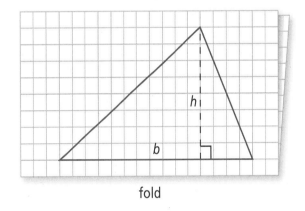

fold

 c. Cut out the triangle so that you end up with two identical triangles. Form a quadrilateral whose area you know. What type of quadrilateral is it? Explain how you *know* it is this type.

 d. Use your results to write a formula for the area of a triangle. Then use your formula to find the exact are of your triangle. Compare this area with your estimate in part (b).

 Area = _____ Formula

 Area = _____ Exact Area

4.2 Areas of Triangles (continued)

3 **ACTIVITY:** Estimating and Finding the Area of a Triangle

Work with a partner. Each grid square represents 1 square centimeter.

- Use estimation to match each triangle with its area.

- Then check your work by finding the exact area of each triangle.

	Area	*Estimate Match*	*Exact Match*
a.	15 cm²	_____	_____
b.	20 cm²	_____	_____
c.	9 cm²	_____	_____
d.	12 cm²	_____	_____
e.	60 cm²	_____	_____
f.	$12\frac{1}{2}$ cm²	_____	_____
g.	$24\frac{1}{2}$ cm²	_____	_____
h.	8 cm²	_____	_____

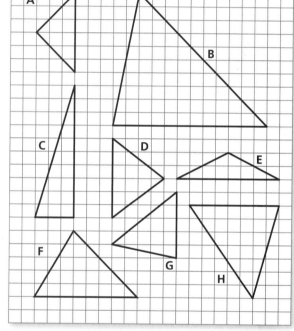

Not drawn to scale

4.2 **Areas of Triangles** (continued)

What Is Your Answer?

4. **PARTNER ACTIVITY** Use the centimeter grid paper to create your own "triangle matching activity." Trade with your partner and solve each other's matching activity.

5. **IN YOUR OWN WORDS** How can you derive a formula for the area of a triangle?

Name_____ Date _____

4.2 Practice
For use after Lesson 4.2

Find the area of the triangle.

1.

5 in.

10 in.

2.

18 ft

18 ft

3.

14 m

11 m

4.

26 cm

9 cm

5.

2 yd

7 yd

6.

13 mm

6 mm

7. A triangular bookend has a base of 4 inches and a height of 8 inches. Find the area of the bookend.

4.3 Areas of Trapezoids
For use with Activity 4.3

Essential Question How can you derive a formula for the area of a trapezoid?

1 **ACTIVITY:** Deriving the Area Formula of a Trapezoid

Work with a partner. Use a piece of centimeter grid paper.

a. Draw *any* trapezoid so that its base lies on one of the horizontal lines of the paper.

b. Estimate the area of your trapezoid (in square centimeters) by counting unit squares.

Area ≈ _____ Estimate

c. Label the height and the bases *inside* the trapezoid.

d. Cut out the trapezoid. Mark the midpoint of the side opposite the height. Draw a line from the midpoint to the opposite upper vertex.

e. Cut along the line. You will end up with a triangle and a quadrilateral. Arrange these two figures to form a figure whose area you know.

f. Use your result to write a *formula* for the area of a trapezoid.

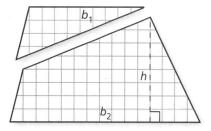

Area = _____ Formula

g. Use your formula to find the area of your trapezoid (in square centimeters).

Area = _____ Exact Area

h. Compare this area with your estimate in part (b).

4.3 **Areas of Trapezoids** (continued)

2 **ACTIVITY:** Writing a Math Lesson

Work with a partner. Use your results from Activity 1 to write a lesson on finding the area of a trapezoid.

Area of a Trapezoid

Key Idea   Use the following steps to find the area of a trapezoid.

   1.

   2.

   3.

> Describe steps you can use to find the area of a trapezoid.

Examples ← Write 2 examples for finding the area of a trapezoid. Include a drawing for each.

   a.   b.

Exercises ← Write 2 exercises for finding the area of a trapezoid. Include an answer sheet.

Find the area.

   1.   2.

4.3 **Areas of Trapezoids** (continued)

What Is Your Answer?

3. **IN YOUR OWN WORDS** How can you derive a formula for the area of
a trapezoid?

4. In this chapter, you used deductive reasoning to derive new area formulas
from area formulas you have already learned. Describe a real-life career in
which deductive reasoning is important.

Name _____ Date _____

4.3 Practice
For use after Lesson 4.3

Find the area of the trapezoid.

1.

15 m

7 m

7 m

2.

9 ft

4 ft

12 ft

Find the area of the trapezoid with height h and bases b_1 and b_2.

3. $h = 10$ yd

$b_1 = 17$ yd

$b_2 = 21$ yd

4. $h = 9$ cm

$b_1 = 4.5$ cm

$b_2 = 5.5$ cm

5. The triangle and the trapezoid have the same area. Base b_2 is twice the length of base b_1. What are the lengths of the bases of the trapezoid?

6 cm

21 cm

b_1

6 cm

b_2

Extension 4.3 **Practice**
For use after Extension 4.3

Find the area of the shaded figure.

1.

2.

3.

4.

Extension 4.3 **Practice** (continued)

Find the area of the figure.

5.

8 cm

4 cm

4 cm

6.

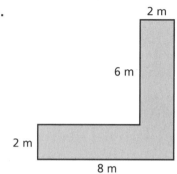

2 m

6 m

2 m

8 m

7.

4 ft

4 ft

5 ft

5 ft

8 ft

8.

5 in.

4 in.

5 in.

6 in.

6 in.

9. You add a 4-foot-by-4-foot section of land to a 6-foot-by-8-foot garden. Find the area of the new garden.

Name_____ Date_____

Essential Question How can you find the lengths of line segments in a coordinate plane?

1 ACTIVITY: Finding Distances on a Map

Work with a partner. The coordinate grid shows a portion of a city. Each square on the grid represents one square mile.

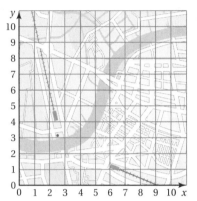

a. A public library is located at (4, 5). City Hall is located at (7, 5). Plot and label these points.

b. How far is the public library from City Hall?

c. A stadium is located 4 miles from the public library. Give the coordinates of several possible locations of the stadium. Justify your answers by graphing.

d. Connect the three locations of the public library, City Hall, and the stadium using your answers in part (c). What shapes are formed?

2 ACTIVITY: Graphing Polygons

Work with a partner. Plot and label each set of points in the coordinate plane. Then connect each set of points to form a polygon.

Rectangle: $A(2, 3)$, $B(2, 10)$, $C(6, 10)$, $D(6, 3)$

Triangle: $E(8, 3)$, $F(14, 8)$, $G(14, 3)$

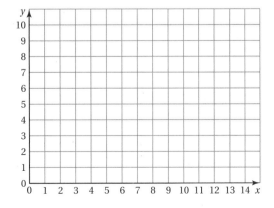

4.4 **Polygons in the Coordinate Plane** (continued)

3 **ACTIVITY:** Finding Distances in a Coordinate Plane

Work with a partner.

a. Find the length of each horizontal line segment in Activity 2.

b. **STRUCTURE** What relationship do you notice between the lengths of the line segments in part (a) and the coordinates of their endpoints? Explain.

c. Find the length of each vertical line segment in Activity 2.

d. **STRUCTURE** What relationship do you notice between the lengths of the line segments in part (c) and the coordinates of their endpoints? Explain.

e. Plot and label the points below in the coordinate plane. Then connect each pair of points with a line segment. Use the relationships you discovered in parts (b) and (d) above to find the length of each line segment. Show your work.

$S(3, 1)$ and $T(14, 1)$

$U(9, 8)$ and $V(9, 0)$

$W(0, 7)$ and $X(0, 10)$

$Y(1, 9)$ and $Z(7, 9)$

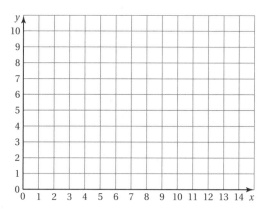

f. Check your answers in part (e) by counting grid lines.

4.4 **Polygons in the Coordinate Plane** (continued)

What Is Your Answer?

4. **IN YOUR OWN WORDS** How can you find the lengths of line segments in a coordinate plane? Give examples to support your explanation.

5. Do the methods you used in Activity 3 work for diagonal line segments? Explain why or why not.

6. Use the Internet or some other reference to find an example of how "finding distances in a coordinate plane" is helpful in each of the following careers.

   a. Archaeologist

   b. Surveyor

   c. Pilot

Name _____ Date _____

Plot and label each pair of points in the coordinate plane. Find the area of the polygon.

1. $A(2, 2), B(2, 6), C(5, 2)$

2. $D(3, 2), E(3, 7), F(6, 2), G(6, 7)$

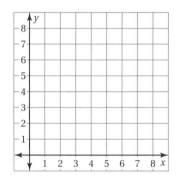

3. $H(3, 3), I(3, 7), J(7, 7), K(7, 3)$

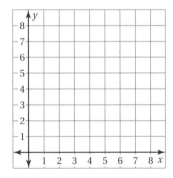

4. $L(1, 2), M(3, 5), N(5, 5), O(7, 2)$

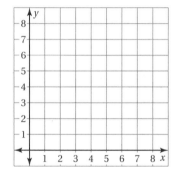

5. The vertices of a sandbox are $P(12, 14), Q(12, 17), R(16, 17)$, and $S(16, 14)$. The coordinates are measured in feet. What is the perimeter of the sandbox?

Name_____ Date_____

Chapter 5 **Fair Game Review**

Using the numbers from the table, find and state the rule in words.

1.

x	y
1	4
2	5
3	6
4	7

2.

x	y
2	6
4	12
6	18
8	24

3.

x	y
12	2
24	14
36	26
48	38

4.

x	y
4	2
5	$\frac{5}{2}$
6	3
7	$\frac{7}{2}$

5. The table shows the results of buying pretzels from a vending machine. The x column is the amount you put into the machine. The y column is the change you receive back from the machine. Complete the table and state the rule in words.

x	y
0.65	0
0.70	0.05
0.75	0.10
1.00	

Chapter 5 **Fair Game Review** (continued)

Evaluate the expression.

6. $\dfrac{5}{9} \cdot \dfrac{1}{3}$

7. $\dfrac{8}{15} \cdot \dfrac{3}{4}$

8. $\dfrac{1}{8} \cdot \dfrac{1}{9}$

9. $\dfrac{2}{3} \div \dfrac{9}{10}$

10. $\dfrac{7}{8} \div \dfrac{11}{16}$

11. $\dfrac{3}{10} \div \dfrac{2}{5}$

12. You have 8 cups of flour. A recipe calls for $\dfrac{2}{3}$ cup of flour. Another recipe calls for $\dfrac{1}{4}$ cup of flour. How much flour do you have left after making the recipes?

Name_____ Date _____

5.1 Ratios
For use with Activity 5.1

Essential Question How can you represent a relationship between two quantities?

---

1 ACTIVITY: Comparing Quantities

Work with a partner. Use the collection of objects to complete each statement.

There are _____ graphing calculators to _____ protractors.

There are _____ protractors to _____ graphing calculators.

There are _____ compasses to _____ protractors.

There are _____ graphing calculators to _____ compasses.

There are _____ protractors to _____ total objects.

The number of graphing calculators is _____ of the total number of objects.

---

2 ACTIVITY: Playing Garbage Basketball

Work with a partner.

- **Take turns shooting a ball or other object into a wastebasket from a reasonable distance.**

- **Organize the numbers of shots you made and shots you missed in a chart.**

5.1 **Ratios** (continued)

a. Write a statement similar to those in Activity 1 that describes the relationship between the number of shots you made and the number of shots you missed.

b. Write a statement similar to those in Activity 1 that describes the relationship between the number of shots you made and the total number of shots.

c. What fraction of your shots did you make? What fraction did you miss?

3 **ACTIVITY:** Reading a Diagram

Work with a partner. You mix different amounts of paint to create new colors. Write a statement that describes the relationship between the amounts of paint shown in each diagram.

a. Blue

Green

There are _____ parts blue for

every _____ parts green.

b. Orange

Yellow

There are _____ for

every _____ .

c. Red

Blue

d. White

Purple

_____   _____

_____   _____

_____   _____

5.1 **Ratios** (continued)

4 **ACTIVITY:** Describing Relationships

Work with a partner. Use a table or a diagram to represent the relationship between the two quantities.

  a. For every 3 boys standing in a line, there are 4 girls.

  b. For each vote Brian received, Sasha received 6 votes.

  c. A class counts the number of vehicles that pass by its school from 1:00 to 2:00 P.M. There are 3 times as many cars as trucks.

  d. A hand sanitizer contains 5 parts aloe for every 2 parts distilled water.

What Is Your Answer?

  5. IN YOUR OWN WORDS How can you represent a relationship between two quantities? Give examples to support your explanation.

  6. MODELING You make 48 pints of pink paint by using 5 pints of red paint for every 3 pints of white paint. Use a diagram to find the number of pints of red paint and white paint in your mixture. Explain.

Name _____ Date _____

Write the ratio. Explain what the ratio means.

1. forks to spoons

2. toothbrushes : toothpaste

Use the table to write the ratio. Explain what the ratio means.

Marble	Number
Blue	8
Red	4
Purple	6

3. red to purple

4. blue to red

5. purple : marbles

6. marbles : blue

7. There are 22 events at an indoor track and field meet. The ratio of track events to field events is 8 : 3. How many of the events are track events?

5.2 Ratio Tables
For use with Activity 5.2

Essential Question How can you find two ratios that describe the same relationship?

---

1 ACTIVITY: Making a Mixture

Work with a partner. A mixture calls for 1 cup of lemonade and 3 cups of iced tea.

a. How many total cups does the mixture contain? _____ cups

For every _____ cup of lemonade, there are _____ cups of iced tea.

b. How do you make a larger batch of this mixture? Describe your procedure and use the table below to organize your results. Add more columns to the table if needed.

Cups of Lemonade						
Cups of Iced Tea						
Total Cups						

c. Which operations did you use to complete your table? Do you think there is more than one way to complete the table? Explain.

d. How many total cups are in your final mixture? How many of those cups are lemonade? How many are iced tea? Compare your results with those of other groups in your class.

5.2 **Ratio Tables** (continued)

e. Suppose you take a sip from every group's final mixture. Do you think all the mixtures should taste the same? Do you think the color of all the mixtures should be the same? Explain your reasoning.

f. Why do you think it is useful to use a table when organizing your results in this activity? Explain.

2 ACTIVITY: Using a Multiplication Table

Work with a partner. Use the information in Activity 1 and the multiplication table below.

	1	2	3	4	5	6	7	8	9	10	11	12
1	1	2	3	4	5	6	7	8	9	10	11	12
2	2	4	6	8	10	12	14	16	18	20	22	24
3	3	6	9	12	15	18	21	24	27	30	33	36
4	4	8	12	16	20	24	28	32	26	40	44	48

a. A mixture contains 8 cups of lemonade. How many cups of iced tea are in the mixture?

b. A mixture contains 21 cups of iced tea. How many cups of lemonade are in the mixture?

c. A mixture has a total of 40 cups. How many cups are lemonade? How many cups are iced tea?

d. Explain how a multiplication table may have helped you in Activity 1.

Name_____ Date_____

3 **ACTIVITY:** Using More than One Ratio to Describe a Quantity

Work with a partner.

a. Find the ratio of pitchers of lemonade to pitchers of iced tea.

b. How can you divide the pitchers into equal groups? Is there more than one way? Use your results to describe the entire collection of pitchers.

c. Three more pitchers of lemonade are added. Is there more than one way to divide the pitchers into equal groups? Explain.

d. The number of pitchers of lemonade and iced tea are doubled. Can you use the ratio in part (b) to describe the entire collection of pitchers? Explain.

What Is Your Answer?

4. **IN YOUR OWN WORDS** How can you find two ratios that describe the same relationship? Give examples to support your explanation.

Name _____ Date _____

Find the missing value(s) in the ratio table. Then write the equivalent ratios.

1.

Kids	3	
Adults	1	3

2.

Basketballs	5	
Footballs	10	20

3.

Apples	4	16
Oranges	5	

4.

CDs		30
DVDs	9	27

5.

Regular	2		32
Decaf	3	12	

6.

Scooters	1	5	
Bikes		15	75

7. You read 1 chapter every hour. You read for 3 hours after school. How many chapters did you read?

Name_____ Date _____

5.3 Rates
For use with Activity 5.3

Essential Question How can you use rates to describe changes in real-life problems?

---

1 ACTIVITY: Stories Without Words

Work with a partner. Each diagram shows a story problem.

- **Describe the story problem in your own words.**

- **Write the rate indicated by the diagram. What are the units?**

a.

b.

c.

5.3 **Rates** (continued)

d.

January 2008
Length: 3 ft

January 2012
Length: 7 ft

2 **ACTIVITY:** Finding Equivalent Rates

Work with a partner. Use the diagrams in Activity 1. Explain how you found each answer.

a. How many miles does the car travel in 1 hour?

b. How much money does the person earn every hour?

c. How much does the population of Sunny Acres Condos increase each year?

d. How many feet does the alligator grow per year?

5.3 **Rates** (continued)

3 **ACTIVITY:** Using a Double Number Line

Work with a partner. Count the number of times you can clap your hands in 12 seconds. Have your partner keep track of the time and record your results.

  a. Use the results to complete the double number line.

Number of claps
0

0   3   6   9   12
Seconds

  b. Explain how to use the double number line to find the number of times you clap your hands in 6 seconds and in 4 seconds.

  c. Find the number of times you can clap your hands in 1 minute. Explain how you found your answer.

  d. How can you find the number of times you can clap your hands in 2 minutes? 3 minutes? Explain.

What Is Your Answer?

  4. **IN YOUR OWN WORDS** How can you use rates to describe changes in real-life problems? Give examples to support your explanation.

  5. **MODELING** Use a double number line to model each story in Activity 1. Show how to use the double number line to answer each question in Activity 2. Why is a double number line a good problem-solving tool for these types of problems?

Name _____ Date _____

5.3 Practice
For use after Lesson 5.3

Write a rate that represents the situation.

1. Calories

0	25	50	75	100

0	5	10	15	20
Minutes

2. Dollars

0	3	6	9	12

0	1	2	3	4
Boxes

Write a unit rate for the situation.

3. 9 strikes in 3 innings

4. 117 points in 13 minutes

Decide whether the rates are equivalent.

5. 30 beats per 20 seconds,
   90 beats per 60 seconds

6. 15 pages in 20 minutes,
   10 pages in 15 minutes

7. One of the valves on the Hoover Dam releases 40,000 gallons of water per second. What is the rate in gallons per minute?

5.4 Comparing and Graphing Ratios
For use with Activity 5.4

Essential Question How can you compare two ratios?

1) ACTIVITY: Comparing Ratio Tables

Work with a partner.

- You make colored frosting by adding 3 drops of red food coloring for every 1 drop of blue food coloring.

- Your teacher makes colored frosting by adding 5 drops of red food coloring for every 3 drops of blue food coloring.

a. Complete the ratio table for each frosting mixture.

Your Frosting	
Drops of Blue	Drops of Red
1	
2	
3	
4	
5	

Your Teacher's Frosting	
Drops of Blue	Drops of Red
3	
6	
9	
12	
15	

b. Whose frosting is bluer? Whose frosting is redder? Justify your answers.

c. **STRUCTURE** Insert and complete a new column for each ratio table above that shows the total number of drops. How can you use this column to answer part (b)?

Name _____ Date _____

2   **ACTIVITY:** Graphing from a Ratio Table

Work with a partner.

a. Explain how you can use the values from the ratio table for your frosting to create a graph in the coordinate plane.

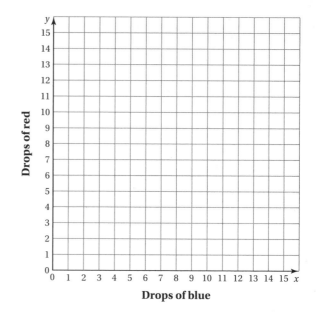

b. Use the values in the table to plot the points. Then connect the points and describe the graph. What do you notice?

c. What does the line represent?

3   **ACTIVITY:** Comparing Graphs From Ratio Tables

Work with a partner. The graph on the next page shows the values from the ratio table for your teacher's frosting.

a. Complete the table and the graph on the next page.

b. Explain the relationship between the entries in the ratio table and the points on the graph.

Your Teacher's Frosting	
Drops of Blue	Drops of Red
3	
6	
9	
12	
15	

5.4 Comparing and Graphing Ratios (continued)

c. How is this graph similar to the graph in Activity 2? How is it different?

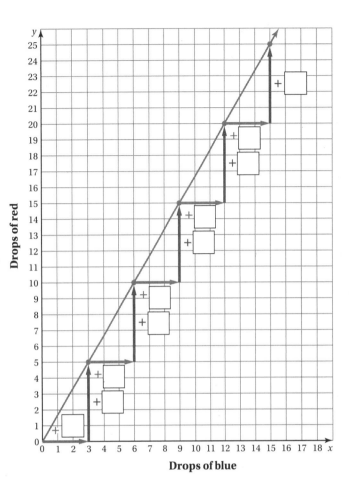

d. How can you use the graphs to determine whose frosting has more red or blue in it? Explain.

What Is Your Answer?

4. IN YOUR OWN WORDS How can you compare two ratios?

5. PRECISION Your teacher's frosting mixture has 7 drops of blue in it. How can you use the graph to find how many drops of red are needed to make the frosting? Is your answer exact? Explain.

Name _____ Date _____

5.4 Practice
For use after Lesson 5.4

Determine which is the better buy.

1.
Iced Tea	A	B
Cost (dollars)	3	4
Refills	2	3

2.
Lunch Meat	A	B
Cost (dollars)	10	6
Pounds	2	1

3.
Movie Rental	A	B
Cost (dollars)	12	11
Rentals	4	3

4.
Buns	A	B
Cost (dollars)	6	5
Packages	3	2

5.
CDs	A	B
Cost (dollars)	60	72
CDs	6	8

6.
Flash Drive	A	B
Cost (dollars)	18	35
Flash Drives	3	5

7. You are making cookies. One recipe calls for 4 cups of chocolate morsels for 3 batches of cookies. A second recipe calls for 5 cups of chocolate morsels for 4 batches of cookies. Which cookies will contain more chocolate morsels?

5.5 Percents
For use with Activity 5.5

Essential Question What is the connection between ratios, fractions, and percents?

1 ACTIVITY: Writing Ratios

Work with a partner.

- Write the fraction of the squares that are shaded.

- Write the ratio of the number of shaded squares to the total number of squares.

- How are the ratios and the fraction related?

- When can you write ratios as fractions?

a.   b.   c.

2 ACTIVITY: Writing Percents as Fractions

Work with a partner.

- What percent of each diagram in Activity 1 is shaded?

5.5 **Percents** (continued)

- What percent of each diagram below is shaded? Write each percent as a fraction in simplest form.

a.

b.

c.

---

3 **ACTIVITY:** Writing Fractions as Percents

Work with a partner. Draw a model to represent the fraction. How can you write the fraction as a percent?

a. $\dfrac{2}{5}$

b. $\dfrac{7}{10}$

c. $\dfrac{3}{5}$

5.5 **Percents** (continued)

d. $\dfrac{3}{4}$ e. $\dfrac{3}{25}$

What Is Your Answer?

4. **IN YOUR OWN WORDS** What is the connection between ratios, fractions, and percents? Give an example with your answer.

5. **REASONING** Your score on a test is 110%. What does this mean?

Name_____ Date _____

Use a 10-by-10 grid to model the percent.

1. 43%

2. 89%

Write the percent as a fraction or mixed number in simplest form.

3. 55%   **4.** 140%   **5.** 12.5%   **6.** 0.6%

Write the fraction or mixed number as a percent.

7. $\frac{3}{4}$ **8.** $\frac{23}{40}$ **9.** $\frac{53}{200}$ **10.** $8\frac{9}{25}$

11. You answered 85% of the questions on the quiz correctly. What fraction of the questions did you answer correctly?

5.6 Solving Percent Problems
For use with Activity 5.6

Essential Question How can you use mental math to find the percent of a number?

"I have a secret way for finding 21% of 80."   "10% is 8, and 1% is 0.8."   "So, 21% is 8 + 8 + 0.8 = 16.8."

---

1 ACTIVITY: Finding 10% of a Number

Work with a partner.

a. How did Newton know that 10% of 80 is 8?

Write 10% as a fraction.

Method 1: Use a model.

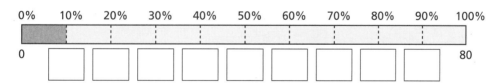

Method 2: Use multiplication.

b. How do you move the decimal point to find 10% of a number?

5.6 **Solving Percent Problems** (continued)

2 **ACTIVITY:** Finding 1% of a Number

Work with a partner.

    a. How did Newton know that 1% of 80 is 0.8?

    b. How do you move the decimal point to find 1% of a number?

3 **ACTIVITY:** Using Mental Math

Work with a partner. Use mental math to find each percent of a number.

    a. 12% of 40   **b.** 19% of 50

4 **ACTIVITY:** Using Mental Math

Work with a partner. Use mental math to find each percent of a number.

    a. 20% tip for a $30 meal   **b.** 18% tip for a $30 meal

    c. 6% sales tax on a $20 shirt   **d.** 9% sales tax on a $20 shirt

5.6 Solving Percent Problems (continued)

e. 6% service charge for
a $200 boxing ticket

f. 2% delivery fee for
a $200 boxing ticket

g. 21% bonus on a total
of 40,000 points

h. 38% bonus on a total
of 80,000 points

What Is Your Answer?

5. **IN YOUR OWN WORDS** How can you use mental math to find the percent of a number?

6. Describe two real-life examples of finding a percent of a number.

7. How can you use 10% of a number to find 20% of the number? 30%? Explain your reasoning.

Name _____ Date _____

5.6 Practice
For use after Lesson 5.6

Find the percent of the number.

1. 30% of 50

2. 12% of 85

3. 2% of 96

4. 150% of 66

5. 7% of 120

6. 3% of 15

Complete the statement using <, >, or =.

7. 70% of 80 _____ 80% of 70

8. 92% of 30 _____ 48% of 75

9. You make homemade lip balm. About 11% of the lip balm is made from beeswax. You make $4\frac{2}{5}$ teaspoons of the lip balm. About how many teaspoons of beeswax do you need? Round your answer to the nearest tenth.

5.7 Converting Measures
For use with Activity 5.7

Essential Question How can you compare lengths between the customary and metric systems?

1 **ACTIVITY:** Customary Measure History

Work with a partner.

a. Match the measure of length with its historical beginning.

Length	Historical Beginning
Inch	The length of a human foot
Foot	The width of a human thumb
Yard	The distance a human can walk in 1000 paces (1 pace = 2 steps)
Mile	The distance from a human nose to the end of an outstretched human arm

b. Use a ruler to measure your thumb, arm, and foot. How do your measurements compare to your answers from part (a)? Are they close to the historical measures?

You already know how to convert measures within the customary and metric systems.

Equivalent Customary Lengths

1 ft = 12 in.   1 yd = 3 ft   1 mi = 5280 ft

Equivalent Metric Lengths

1 m = 1000 mm   1 m = 100 cm   1 km = 1000 m

You will learn how to convert between the two systems.

Converting Between Systems

1 in. = 2.54 cm

1 mi ≈ 1.61 km

5.7 Converting Measures (continued)

2 ACTIVITY: Comparing Measures

Work with a partner. Answer each question. Explain your answer. Use a diagram in your explanation.

	Metric	*Customary*
a. Car Speed: Which is faster?	80 km/h	60 mi/h
b. Trip Distance: Which is farther?	200 km	200 mi
c. Human Height: Who is taller?	180 cm	5 ft 8 in.
d. Wrench Width: Which is wider?	8 mm	5/16 in.
e. Swimming Pool Depth: Which is deeper?	1.4 m	4 ft

Name_____ Date_____

3 **ACTIVITY:** Changing Units in a Rate

Work with a partner. Change the units of the rate by multiplying by a "Magic One." Write your answer as a unit rate. Show your work.

Original Rate	*Magic One*	*New Units*	*Unit Rate*

a. $\dfrac{\$120}{h}$ $\times$ $\dfrac{\boxed{}}{\boxed{}}$ $=$ $\dfrac{\boxed{}}{\boxed{}}$ $=$ $\dfrac{\$\boxed{}}{1 \text{ min}}$

b. $\dfrac{\$3}{\min}$ $\times$ $\dfrac{\boxed{}}{\boxed{}}$ $=$ $\dfrac{\boxed{}}{\boxed{}}$ $=$ $\dfrac{\$\boxed{}}{1 \text{ h}}$

c. $\dfrac{12 \text{ in.}}{ft}$ $\times$ $\dfrac{\boxed{}}{\boxed{}}$ $=$ $\dfrac{\boxed{}}{\boxed{}}$ $=$ $\dfrac{\boxed{} \text{ in.}}{1 \text{ yd}}$

d. $\dfrac{2 \text{ ft}}{\text{week}}$ $\times$ $\dfrac{\boxed{}}{\boxed{}}$ $=$ $\dfrac{\boxed{}}{\boxed{}}$ $=$ $\dfrac{\boxed{} \text{ ft}}{1 \text{ yr}}$

What Is Your Answer?

4. One problem-solving strategy is called *Working Backwards*. What does this mean? How can you use this strategy to find the rates in Activity 3?

5. IN YOUR OWN WORDS How can you compare lengths between the customary and the metric systems? Give examples with your description.

Name _____ Date _____

5.7 Practice

For use after Lesson 5.7

Complete the statement.

1. $3 \text{ m} \approx$ _____ ft

2. $32 \text{ cm} \approx$ _____ in.

3. $16 \text{ qt} \approx$ _____ L

4. $\dfrac{50 \text{ mi}}{\text{h}} \approx \dfrac{\boxed{} \text{ km}}{\text{h}}$

5. $\dfrac{25 \text{ gal}}{\text{min}} = \dfrac{\boxed{} \text{ qt}}{\text{sec}}$

6. $\dfrac{1000 \text{ m}}{\text{sec}} = \dfrac{\boxed{} \text{ km}}{\text{min}}$

7. Your doctor prescribes you to take 400 milligrams of medicine every 8 hours. How many ounces of medicine do you take in a day?

Chapter 6 Fair Game Review

Use a number line to order the numbers from least to greatest.

1. 0.2, 0.54, 0.61, 0.4

2. 0.3, 0.45, 0.11, 0.02

3. 1.7, 1.24, 1.02, 1.33

4. 1.01, 1.42, 1.06, 1.2

5. 0.98, 1.23, 0.87, 0.9

6. 1.4, 0.06, 1.23, 0.5

7. 0.003, 0.03, 0.033, 0.031

8. 0.02, 0.002, 0.2, 0.022

9. In your class, 0.58 of the students bring a piece of whole fruit for a snack and 0.36 of the students bring a snack pack of crackers. Which group of students brings in more food items for a snack?

Chapter 6 Fair Game Review (continued)

Complete the number sentence with <, >, or =.

10. 5 ___ 8

11. 13 ___ 9

12. 0.3 ___ $\frac{3}{8}$

13. 0.68 ___ $\frac{17}{25}$

14. 3.6 ___ $\frac{12}{5}$

15. 0.06 ___ 0.062

Find three numbers that make the number sentence true.

16. 0.35 < ___

17. $\frac{4}{9}$ ≥ ___

18. $2\frac{3}{5}$ ≤ ___

19. $\frac{1}{10}$ < ___

20. 0.485 ≥ ___

21. 5.87 ≤ ___

22. During a trivia game, you answered 18 out of 25 questions correctly. Your friend answered 0.7 of the questions correctly. Write a number sentence for who had the greater number of correct answers.

6.1 Integers
For use with Activity 6.1

Essential Question How can you represent numbers that are less than 0?

---

1 ACTIVITY: Reading Thermometers

Work with a partner. The thermometers show the temperatures in four cities.

> *Honolulu, Hawaii*   *Anchorage, Alaska*
> *Death Valley, California*   *Seattle, Washington*

Write each temperature. Then match each temperature with its most appropriate location.

a.   b.   c.   d.

   e. How would you describe all the temperatures in relation to 0°F?

---

2 ACTIVITY: Describing a Temperature

Work with a partner. The thermometer on the next page shows the coldest temperature ever recorded in Seattle, Washington.

   a. What is the temperature?

6.1 **Integers** (continued)

b. How do you write temperatures that are colder than this?

c. Suppose the record for the coldest temperature in Seattle is broken by 10 degrees. What is the new coldest temperature? Draw a thermometer that shows the new coldest temperature.

d. How is the new coldest temperature different from the temperatures in Activity 1?

3 **ACTIVITY:** Extending the System of Whole Numbers

Work with a partner.

a. Draw a number like the one shown on a sheet of paper. Complete the number line using whole numbers only.

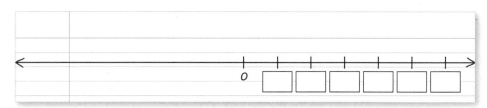

b. Fold the paper with your number line around 0 so that the lines overlap. Make tick marks on the other side of the number line to match the tick marks for the whole numbers.

c. **STRUCTURE** Compare this number line to the thermometers from Activities 1 and 2. What do you think the new tick marks represent? How would you label them?

6.1 Integers (continued)

What Is Your Answer?

4. **IN YOUR OWN WORDS** How can you represent numbers that are less than 0?

5. Describe another real-life example that uses numbers that are less than 0?

6. **REASONING** How are the temperatures shown by the thermometers at the right similar? How are they different?

7. **WRITING** The temperature in a town on Thursday evening is 25°F. On Sunday morning, the temperature drops below 0°F. Write a story to describe what may have happened in the town. Be sure to include the temperatures for each day.

6.1 Practice
For use after Lesson 6.1

Write a positive or negative integer that represents the situation.

1. You gain 60 points in a board game.

2. The temperature is 9 degrees below zero.

3. The stock market drops 18 points.

4. You earn $125 at your job.

Graph the integer and its opposite.

5. -9

6. 5

7. 14

8. -20

9. -56

10. 850

11. You hike 72 feet up a mountain. The next day, you hike 12 feet down the mountain. Write an integer to represent each situation.

6.2 Comparing and Ordering Integers
For use with Activity 6.2

Essential Question How can you use a number line to order real-life events?

---

1 ACTIVITY: Seconds to Takeoff

Work with a partner. You are listening to a command center before the liftoff of a rocket.

You hear the following:

"**T minus 10 seconds…go for main engine start…T minus 9…8… 7…6…5…4…3…2…1…we have liftoff.**"

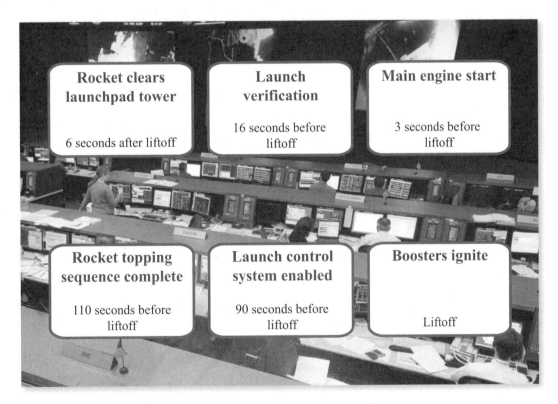

Rocket clears launchpad tower	Launch verification	Main engine start
6 seconds after liftoff	16 seconds before liftoff	3 seconds before liftoff

Rocket topping sequence complete	Launch control system enabled	Boosters ignite
110 seconds before liftoff	90 seconds before liftoff	Liftoff

a. Draw a number line. Then locate the events shown above at appropriate points on the number line.

b. Which event occurs at zero on your number line? Explain.

6.2 **Comparing and Ordering Integers** (continued)

c. Which of the events occurs first? Which of the events occurs last? How do you know?

d. List the events in the order they occurred.

2 ACTIVITY: Being Careful with Terminology

Work with a partner.

a. Use a number line to show that the phrase "3 seconds away from liftoff" can have two meanings.

b. Reword the phrase "3 seconds away from liftoff" in two ways so that each meaning is absolutely clear.

c. Explain why you must be very careful with terminology if you are working in the command center for a rocket launch.

6.2 **Comparing and Ordering Integers** (continued)

3 **ACTIVITY:** A Day in the Life of an Astronaut

Make a time line that shows a day in the life of an astronaut. Use the Internet or another reference source to gather information.

- Use a number line with units representing hours. Start at 12 hours before liftoff and end at 12 hours after liftoff. Locate the liftoff at 0. Assume liftoff occurs at noon.

- Include at least five events before liftoff, such as when the astronauts suit up.

- Include at least five events after liftoff, such as when the rocket enters Earth's orbit.

- How do you determine where each event occurs on the number line?

What Is Your Answer?

4. **IN YOUR OWN WORDS** How can you use a number line to order real-life events?

5. Describe how you can use a number line to create a time line.

6.2 Practice
For use after Lesson 6.2

Complete the statement using < or >.

1. 0 _____ 9

2. 8 _____ -3

3. -5 _____ -7

4. -12 _____ 0

5. -6 _____ -2

6. -21 _____ -40

Order the integers from least to greatest.

7. $1, -3, -6, 5, 0$

8. $2, -4, -9, 3, -1$

9. $11, -11, 18, -18, -8$

10. $21, -14, -35, 28, -7$

11. Your miniature golf scores for the first half of a course are $-3, 7, -2, -5, 3,$ $-1, 0, 4, -4$. Order the scores from least to greatest.

6.3 Fractions and Decimals on the Number Line
For use with Activity 6.3

Essential Question How can you use a number line to compare positive and negative fractions and decimals?

---

1 ACTIVITY: Locating Fractions on a Number Line

On your time line for "A Day in the Life of an Astronaut" from Activity 3 in Section 6.2, include the following events. Represent each using a fraction or a mixed number.

a. Radio Transmission: 10:30 A.M.

b. Space Walk: 7:30 P.M.

c. Physical Exam: 4:45 A.M.

d. Photograph Taken: 3:15 A.M.

e. Float in the Cabin: 6:20 P.M.

f. Eat Dinner: 8:40 P.M.

---

2 ACTIVITY: Fractions and Decimals on a Number Line

Work with a partner. Find a number that is between the two numbers. The number must be greater than the number on the left *and* less than the number on the right.

a.

6.3 **Fractions and Decimals on the Number Line** (continued)

b.

c.

---

3 **ACTIVITY:** Decimals on a Number Line

Work with a partner.

Snorkeling: Scuba diving: Deep sea diving:
−5 meters   −50 meters   −700 meters

a. Write the position of each diver in kilometers.

b. CHOOSE TOOLS Would a horizontal or a vertical number line be more appropriate for representing these data? Why?

6.3　**Fractions and Decimals on the Number Line** (continued)

   c. Use a number line to order the positions from deepest to shallowest.

What Is Your Answer?

  4. **IN YOUR OWN WORDS** How can you use a number line to compare positive and negative fractions and decimals?

  5. Draw a number line. Graph and label three values between −2 and −1.

6.3 Practice
For use after Lesson 6.3

Complete the statement using < or >.

1. 3.7 _____ −3.2

2. −10.4 _____ −10.04

3. $-\dfrac{2}{3}$ _____ $\dfrac{1}{4}$

4. $-2\dfrac{2}{3}$ _____ $-2\dfrac{1}{2}$

Order the numbers from least to greatest.

5. −2, 0, −4, 2, 3

6. −6.3, 4.2, 7.7, −3.9, 3.4

7. $1\dfrac{4}{5}, \dfrac{1}{2}, -3\dfrac{7}{8}, -\dfrac{7}{9}, -\dfrac{3}{5}$

8. $-3, 0.6, \dfrac{1}{4}, 0, -1\dfrac{2}{3}$

9. An archaeologist discovers two artifacts. Compare the positions of the artifacts.

6.4 Absolute Value
For use with Activity 6.4

Essential Question How can you describe how far an object is from sea level?

1 ACTIVITY: Sea Level

Work with a partner. Write an integer that represents the elevation of each object. How far is each object from sea level? Explain your reasoning.

a. Boeing 747 ——————→

b. Seaplane ——————→

c. Bald eagle ——————→

d. Leatherback turtle ———→
e. U.S.S. Dolphin ————→

f. Whale ——————→

g. Jason Jr. ——————→

h. Alvin ——————→

i. Kaiko ——————→

5000 meters
4000 meters
3000 meters
2000 meters
1000 meters
0 meters
−1000 meters
−2000 meters
−3000 meters
−4000 meters
−5000 meters
−6000 meters
−7000 meters
−8000 meters

6.4   **Absolute Value** (continued)

2 **ACTIVITY:** Finding a Distance

Work with a partner. Use the diagram in Activity 1.

   a. What integer represents sea level?

   b. The vessel *Kaiko* ascends to the same depth as the U.S.S. *Dolphin*. About how many meters did *Kaiko* travel? Explain how you found your answer.

   c. The vessel *Jason Jr.* descends to the same depth as the *Alvin*. About how many meters did *Jason Jr.* travel? Explain how you found your answer.

   d. **REASONING** Which pairs of objects are the same distance from sea level? How do you know?

   e. **REASONING** An airplane is the same distance from sea level as the *Kaiko*. How far is the airplane from sea level?

Name_____ Date _____

3 **ACTIVITY:** Oceanography Project

Work with a partner. Use the Internet or some other resource to write a report that describes two ways in which mathematics is used in oceanography.

Here are two possible ideas. You can use one or both of these, or you can use other ideas.

Diving Bell

Mine Neutralization Vehicle

What Is Your Answer?

4. **IN YOUR OWN WORDS** How can you describe how far an object is from sea level?

5. **PRECISION** In Activity 1, an object has an elevation of -7500 meters. Is -7500 greater than or less than -7000? Does this object have a depth greater than or less than 7000 meters? Explain your reasoning.

6.4 Practice
For use after Lesson 6.4

Find the absolute value.

1. $|-5|$
2. $|7|$

3. $|0|$
4. $|-31|$

Complete the statement using <, >, or =.

5. 7 _____ $|-4|$
6. -9 _____ $|-10|$
7. $|8|$ _____ $|-8|$

Order the values from least to greatest.

8. $0, -5, |-6|, |-2|, 4$
9. $|-12|, -21, |25|, |-31|, -14, 33$

10. You go to a store in a mall that is on the fourth floor above the main level. Your friend goes to a store that is two floors below the main level.

   a. Write an integer for the position of each person relative to the main level.

   b. Find the absolute value of each integer.

   c. Who is farther from the main level? Explain.

Name_____ Date _____

6.5 The Coordinate Plane
For use with Activity 6.5

Essential Question How can you graph and locate points that contain negative numbers in a coordinate plane?

1 ACTIVITY: Forming the Entire Coordinate Plane

Work with a partner.

a. In the middle of a sheet of grid paper, construct a horizontal number line. Label the tick marks. On a different sheet of grid paper, construct and label a similar vertical number line.

b. Cut out the vertical number line and tape it on top of the horizontal number line so that the zeros overlap. Make sure the number lines are perpendicular to one another. How many regions did you form by doing this?

c. **REASONING** What ordered pair represents the point where the number lines intersect? Why do you think this point is called the *origin*? Explain.

2 ACTIVITY: Describing Points in the Coordinate Plane

Work with a partner. Use your perpendicular number lines from Activity 1.

a. Plot and label (3, 2) in your coordinate plane. Shade this region in your coordinate plane. What do you notice about the integers along the number lines that surround (3, 2)?

b. Can you plot a point in your coordinate plane so that it is surrounded by negative numbers on the axes? If so, where is this point? Use a different color to shade this region in your coordinate plane.

c. What do you notice about the integers along the number lines for points in the regions that are not shaded?

6.5 **The Coordinate Plane** (continued)

d. STRUCTURE Describe how you would plot (−3, −2). How is plotting this
point similar to plotting (3, 2)? Plot (−3, −2) in your coordinate plane.

e. REASONING Where in your coordinate plane do you plot (2, −4)? Where
do you plot (−2, 4)? Explain your reasoning.

---

3 **ACTIVITY:** Plotting Points in a Coordinate Plane

**Work with a partner. Plot and connect the points to make a picture.
Describe and color the picture when you are done.**

1(6, 9)   **2**(4, 11)   **3**(2, 12)   **4**(0, 11)   **5**(−2, 9)

6(−6, 2)   **7**(−9, 1)   **8**(−11, −3)   **9**(−7, 0)   **10**(−5, −1)

11(−5, −5)   **12**(−4, −8)   **13**(−6, −10)   **14**(−3, −9)   **15**(−3, −10)

16(−4, −11)   **17**(−4, −12)

18(−3, −11)   **19**(−2, −12)

20(−2, −11)   **21**(−1, −12)

22(−1, −11)   **23**(−2, −10)

24(−2, −9)   **25**(1, −9)

26(2, −8)   **27**(2, −10)

28(1, −11)   **29**(1, −12)

30(2, −11)   **31**(3, −12)

32(3, −11)   **33**(4, −12)

34(4, −11)   **35**(3, −10)

36(3, −8)   **37**(4, −6)

38(6, 0)   **39**(9, −3)

40(9, −1)   **41**(8, 1)

42(5, 3)   **43**(3, 6)

44(3, 7)   **45**(4, 8)

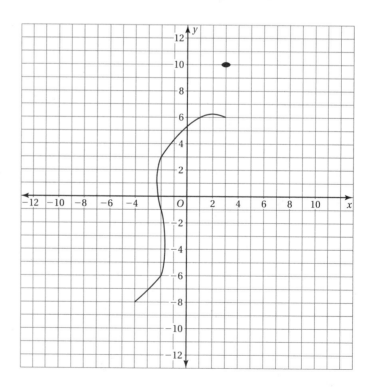

6.5 **The Coordinate Plane** (continued)

What Is Your Answer?

4. **IN YOUR OWN WORDS** How can you graph and locate points that contain negative numbers in a coordinate plane?

5. Make up your own "dot-to-dot" picture. Use at least 20 points. Your picture should have at least two points in each quadrant.

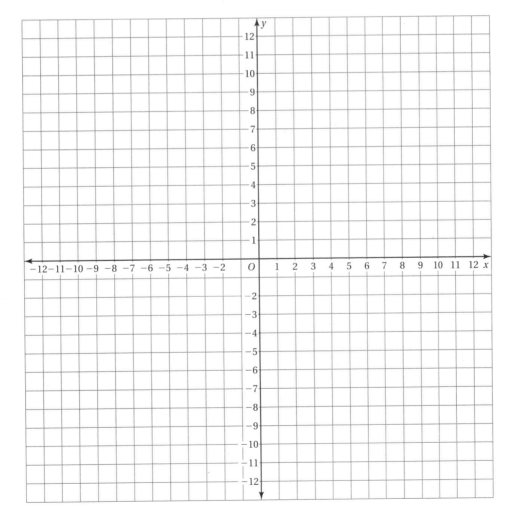

Name _____ Date _____

6.5 **Practice**
For use after Lesson 6.5

Plot the ordered pair in the coordinate plane. Describe the location of the point.

1. $A(8, 4)$

2. $B(-3, 5)$

3. $C(-2, -2)$

4. $D(4, -7)$

5. $E(-6, -5)$

6. $F(-9, 7)$

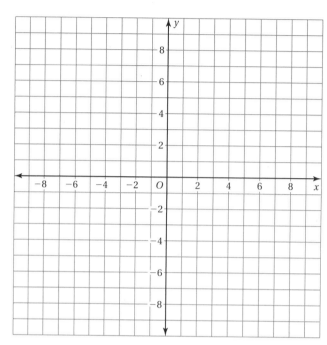

7. The coordinates of three vertices of a rectangle are shown in the figure. What are the coordinates of the fourth vertex?

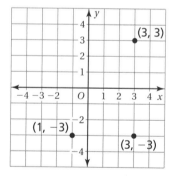

8. Your house is located at $(-4, 3)$, which is 4 blocks west and 3 blocks north of the center of town. To get from your house to the mall, you walk 7 blocks east and 4 blocks south.

   a. What ordered pair corresponds to the location of the mall?

   b. Is your house or the mall closer to the center of town? Explain.

Extension 6.5 **Practice**
For use after Extension 6.5

Reflect the point in (a) the x-axis and (b) the y-axis.

1. $(2, 4)$

2. $(-3, 1)$

3. $(-4, -1)$

4. $(2, -3)$

5. $(0, -2)$

6. $(-1, 0)$

Name _____ Date _____

Extension 6.5 **Practice** (continued)

Reflect the point in the *x*-axis followed by the *y*-axis.

7. (2, 2)

8. (−4, 3)

9. (−1.5, −1.5)

10. (3.5, −3.5)

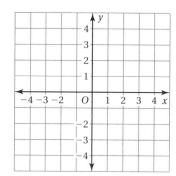

11. The vertices of a triangle are (2, 2), (4, 4), and (4, 2). Reflect the triangle in the *y*-axis. Give the coordinates of the reflected triangle.

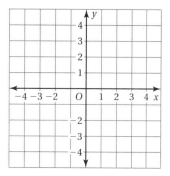

Chapter 7 Fair Game Review

Evaluate the expression when $x = 3$ and $y = 5$.

1. $2xy$

2. $\dfrac{6y}{x}$

3. $4y - x$

4. $y^2 - 7x + 2$

Evaluate the expression when $x = \dfrac{1}{4}$ and $y = 8$.

5. $3xy$

6. $16x + 5y$

7. $\dfrac{y}{2x}$

8. $2(10 - 24x) + y^2$

9. After m months, you paid $25 + 10m$ for your computer. How much did you pay after 6 months?

Chapter 7 **Fair Game Review** (continued)

Write the phrase as an expression.

10. three more than twice a number k

11. half of a number q plus eight

12. a number p decreased by six

13. nine times a number x

14. five divided by a number n

15. one plus the product of a number y and three

16. Each classmate contributes $2 for charity. Write an expression for the amount of money raised by your class.

17. You save half of the money from your paycheck plus an extra six dollars to buy a new bike. Write an expression for the amount of money you save from each paycheck.

7.1 Writing Equations in One Variable
For use with Activity 7.1

Essential Question How does rewriting a word problem help you solve the word problem?

---

1 ACTIVITY: Rewriting a Word Problem

Work with a partner. Read the problem several times. Think about how you could rewrite the problem, leaving out information that you do not need to solve the problem.

Given Problem (63 words)

Your minivan has a flat, rectangular area in the back. When you fold down the rear seats of the van and move them forward, the width of the rectangular area in the van is increased by 2 feet, as shown in the diagram.

By how many square feet does the rectangular area increase when the rear seats are folded down and moved forward?

Rewritten Problem (28 words)

When you fold down the back seats of a minivan, the added area is a 5-foot by 2-foot rectangle. What is the area of this rectangle?

Can you make the problem even simpler?

Added Area = 2 × 5
= 10 ft²

Explain why your rewritten problem is easier to read.

7.1 **Writing Equations in One Variable** (continued)

2 **ACTIVITY:** Rewriting a Word Problem

Work with a partner. Rewrite each problem using fewer words. Leave out information that you do not need to solve the problem. Then solve the problem.

  a. (63 words)

> A supermarket is having its grand opening on Saturday morning. Every fifth customer will receive a $10 coupon for a free turkey. Every seventh customer will receive a $3 coupon for 2 gallons of ice cream. You are the manager of the store and you expect 400 customers. How many of each type of coupon should you plan to give away?

  b. (71 words)

> You and your friend are at a football game. The stadium is 4 miles from your home. You each brought $5 to spend on refreshments. During the third quarter of the game, you say, "I read that the greatest distance that a baseball has been thrown is 445 feet 10 inches." Your friend says, "That's about one and a half times the length of the football field." Is your friend correct?

7.1 **Writing Equations in One Variable** (continued)

c. **(90 words)**

> You are visiting your cousin who lives in the city. To get back home, you take a taxi. The taxi charges $2.10 for the first mile and $0.90 for each additional mile. After riding 13 miles, you decide that the fare is going to be more than the $20 you have with you. So, you tell the driver to stop and let you out. Then you call a friend and ask your friend to come pick you up. After paying the driver, how much of your $20 is left?

What Is Your Answer?

3. **IN YOUR OWN WORDS** How does rewriting a word problem help you solve the word problem? Make up a word problem that has more than 50 words. Then show how you can rewrite the problem using at most 25 words.

7.1 Practice
For use after Lesson 7.1

Write the word sentence as an equation.

1. 27 is 3 times a number y.

2. The difference of a number x and 4 is 3.

3. 8 more than a number p is 17.

4. Half of a number q is 14.

Write an equation that can be used to find the value of x.

5. Perimeter of rectangle: 32 cm

6. Perimeter of triangle: 20 in.

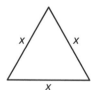

7. You spend $16 on 3 notebooks and x binders. Notebooks cost $2 each and binders cost $5 each. Write an equation you can use to find the number of binders you bought.

Name_____ Date_____

7.2 Solving Equations Using Addition or Subtraction
For use with Activity 7.2

Essential Question How can you use addition or subtraction to solve an equation?

When two sides of a scale weigh the same, the scale will balance.

When you add or subtract the same amount on each side of the scale, it will still balance.

1 ACTIVITY: Solving an Equation

Work with a partner.

a. Use a model to solve $n + 3 = 7$.

- Explain how the model represents the equation $n + 3 = 7$.

- How much does one ⬤ weigh? How do you know?

The solution is $n = $ _____.

b. Describe how you could check your answer in part (a).

7.2   **Solving Equations Using Addition or Subtraction** (continued)

   c. Which model below represents the solution of $n + 1 = 9$? How do you know?

2   **ACTIVITY:** Solving Equations

Work with a partner. Solve the equation using the method in Activity 1.

   a. $n + 5 = 10$ **b.** $x + 2 = 11$

   c. $6 = y + 3$ **d.** $8 = m + 8$

3   **ACTIVITY:** Solving Equations Using Mental Math

Work with a partner. Write a question that represents the equation. Use mental math to answer the question. Then check your solution.

Equation	Question	Solution	Check
a. $x + 1 = 5$			
b. $4 + m = 11$			
c. $8 = a + 3$			
d. $x - 9 = 21$			
e. $13 = p - 4$			

What Is Your Answer?

   4. **REPEATED REASONING** In Activity 3, how are parts (d) and (e) different from parts (a)–(c)? Did your process to find the solution change? Explain.

7.2 **Solving Equations Using Addition or Subtraction** (continued)

5. Decide whether the statement is *true* or *false*. If false, explain your reasoning.

   a. In an equation, you can use any letter as a variable.

   b. The goal in solving an equation is to get the variable by itself.

   c. In the solution, the variable must be on the left side of the equal sign.

   d. If you add a number to one side, you should subtract it from the other side.

6. **IN YOUR OWN WORDS** How can you use addition or subtraction to solve an equation? Give two examples to show how your procedure works.

7. Are the following equations equivalent? Explain your reasoning.

   $$x - 5 = 12 \qquad \text{and} \qquad 12 = x - 5$$

7.2 Practice
For use after Lesson 7.2

Tell whether the given value is a solution of the equation.

1. $34 + x = 46$; $x = 12$

2. $y - 9 = 14$; $y = 22$

3. $6d = 54$; $d = 9$

4. $\dfrac{n}{3} = 13$; $n = 39$

Solve the equation. Check your solution.

5. $7 + k = 11$

6. $p - 24 = 13$

7. $b - 16 = 7$

8. $\dfrac{2}{5} + m = \dfrac{5}{6}$

9. In the heavyweight class of professional wrestling, the junior weight limit is 190 pounds. This is 15 pounds heavier than the light heavyweight limit. Write and solve an equation to find the weight limit of the light heavyweight class.

7.3 Solving Equations Using Multiplication or Division
For use with Activity 7.3

Essential Question How can you use multiplication or division to solve an equation?

1 ACTIVITY: Find Missing Dimensions

Work with a partner. Describe how you would find the value of *x*. Then find the value and check your result.

a. rectangle

Area = 24 square units

6

x

b. parallelogram

Area = 20 square units

x

5

c. triangle

Area = 28 square units

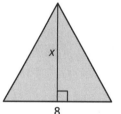

x

8

2 ACTIVITY: Using an Equation to Model a Story

Work with a partner.

a. Use a model to solve the problem.

> Three people go out to lunch. They decide to share the $12 bill evenly. How much does each person pay?

- What equation does the model represent? Explain how this represents the problem.

- How much does one ⬤ weigh? How do you know?

Each person pays _____.

b. Describe how you can check your answer in part (a).

7.3 Solving Equations Using Multiplication or Division (continued)

3 **ACTIVITY:** Using Equations to Model a Story

Work with a partner.

- **What is the unknown?**
- **Write an equation that represents each problem.**
- **What does the variable in your equation represent?**
- **Explain how you can solve the equation.**
- **Answer the question.**

Problem	Equation
a. Three robots go out to lunch. They decide to share the $11.91 bill evenly. How much does each robot pay?	_____
b. On Earth, objects weigh 6 times what they weigh on the Moon. A robot weighs 96 pounds on Earth. What does it weigh on the Moon?	_____
c. At maximum speed, a robot runs 6 feet in 1 second. How many feet does the robot run in 1 minute?	_____
d. Four identical robots lie on the ground head-to-toe and measure 14 feet. How tall is each robot?	_____

7.3 **Solving Equations Using Multiplication or Division** (continued)

What Is Your Answer?

4. Complete each sentence by matching.

- The inverse operation of addition
- The inverse operation of subtraction
- The inverse operation of multiplication
- The inverse operation of division

- is multiplication.
- is subtraction.
- is addition.
- is division.

5. IN YOUR OWN WORDS How can you use multiplication or division to solve an equation? Give two examples to show how your procedure works.

7.3 Practice
For use after Lesson 7.3

Solve the equation. Check your solution.

1. $7k = 77$

2. $\dfrac{p}{5} = 10$

3. $3 = \dfrac{m}{12}$

4. $4a = 36$

5. $5 \bullet x = 12$

6. $4.2 = \dfrac{c}{8}$

7. You earn $5 for every friendship bracelet you sell. Write and solve an equation to find the number of bracelets you have to sell to earn $85.

8. You practice the piano for 30 minutes each day. Write and solve an equation to find the total time t you spend practicing the piano in a week.

7.4 Writing Equations in Two Variables
For use with Activity 7.4

Essential Question How can you write an equation in two variables?

1 ACTIVITY: Writing an Equation in Two Variables

Work with a partner. You earn $8 per hour working part-time at a store.

a. Complete the table.

Hours Worked	Money Earned (dollars)
1	
2	
3	
4	
5	

b. Use the values from the table to complete the graph. Then answer each question below and on the next page.

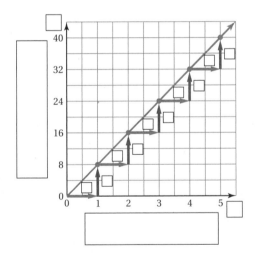

- What does the horizontal axis represent? What variable did you use to identify it?

7.4 **Writing Equations in Two Variables** (continued)

- What does the vertical axis represent? What variable did you use to identify it?

- How are the ordered pairs in the graph related to the values in the table?

- How are the horizontal and vertical distances shown on the graph related to the values in the table?

c. How can you write an equation that shows how the two variables are related?

d. What does the line in the graph represent?

2 ACTIVITY: Describing Variables

Work with a partner. Use the equation you wrote in Activity 1.

a. How is this equation different from the equations earlier in this chapter?

b. One of the variables in this equation *depends* on the other variable. Determine which variable is which by answering the following questions:

- Does the amount of money you earn *depend* on the number of hours you work?

- Does the number of hours you work *depend* on the amount of money you earn?

What do you think is the significance of having two types of variables? How do you think you can use these types of variables in real life?

7.4 **Writing Equations in Two Variables** (continued)

3 **ACTIVITY:** Describing a Formula in Two Variables

Work with a partner. Recall that the perimeter of a square is 4 times its side length.

a. Write the formula for the perimeter of a square. Tell what each variable represents.

b. Describe how the perimeter of a square changes as its side length increases by 1 unit. Use a table and a graph to support your answer.

c. In your formula, which variable depends on which?

What Is Your Answer?

4. **IN YOUR OWN WORDS** How can you write an equation in two variables?

5. The equation $y = 7.75x$ shows how the number of movie tickets is related to the total amount of money spent. Describe what each part of the equation represents.

6. **CHOOSE TOOLS** In Activity 1, you want to know the amount of money you earn after working 30.5 hours during a week. Would you use the table, the graph, or the equation to find your earnings? What are your earnings? Explain your reasoning.

7. Give an example of another real-life situation that you can model by an equation in two variables.

Name _____ Date _____

Tell whether the ordered pair is a solution of the equation.

1. $y = 2x$; $(0, 2)$

2. $y = 6x$; $(2, 12)$

3. $y = 2x + 3$; $(3, 9)$

4. $y = x + 4$; $(1, 3)$

Identify the independent and dependent variables.

5. The equation $p = 8.65h$ gives the amount p (in dollars) of pay a clerk receives for working h hours.

6. The equation $P = 4s$ gives the perimeter P (in inches) of a square mouse pad with a side length of s inches.

7. The equation $c = 42t + 42$ gives the total cost c (in dollars) of a grocery bill with a sales tax of t percent (in decimal form).

8. Avocados cost $3 per pound. Write and graph an equation in two variables that represents the cost of buying avocados.

7.5 Writing and Graphing Inequalities
For use with Activity 7.5

Essential Question How can you use a number line to represent solutions of an inequality?

1 ACTIVITY: Understanding Inequality Statements

Work with a partner. Read the statement. Circle each number that makes the statement true, and then answer the questions.

a. **"Your friend is *more than* 3 minutes late."**

$$-3 \quad -2 \quad -1 \quad 0 \quad 1 \quad 2 \quad 3 \quad 4 \quad 5 \quad 6$$

- What do you notice about the numbers that you circled?

- Is the number 3 included? Why or why not?

- Write four other numbers that make the statement true.

b. **"The temperature is *at most* 2 degrees."**

$$-5 \quad -4 \quad -3 \quad -2 \quad -1 \quad 0 \quad 1 \quad 2 \quad 3 \quad 4$$

- What do you notice about the numbers that you circled?

- Can the temperature be exactly 2 degrees? Explain.

- Write four other numbers that make the statement true.

c. **"You need *at least* 4 pieces of paper for your math homework."**

$$-3 \quad -2 \quad -1 \quad 0 \quad 1 \quad 2 \quad 3 \quad 4 \quad 5 \quad 6$$

- What do you notice about the numbers that you circled?

- Can you have exactly 4 pieces of paper? Explain.

- Write four other numbers that make the statement true.

7.5 **Writing and Graphing Inequalities** (continued)

    d. "After playing a video game for 20 minutes, you have *fewer than* 6 points."

        −2   −1   0   1   2   3   4   5   6   7

- What do you notice about the numbers that you circled?

- Is the number 6 included? Why or why not?

- Write four other numbers that make the statement true.

2 **ACTIVITY:** Understanding Inequality Symbols

Work with a partner.

    a. Consider the statement "x is a number such that $x < 2$."

- Can the number be exactly 2? Explain.

- Circle each number that makes the statement true.

        −5   −4   −1   −2   −1   0   1   2   3   4

- Write four other numbers that make the statement true.

    b. Consider the statement "x is a number such that $x \geq 1$."

- Can the number be exactly 1? Explain.

- Circle each number that makes the statement true.

        −5   −4   −3   −2   −1   0   1   2   3   4

- Write four other numbers that make the statement true.

7.5 **Writing and Graphing Inequalities** (continued)

---

3 **ACTIVITY:** How Close Can You Come to 0?

Work with a partner.

   a. Which number line shows $x > 0$? Which number line shows $x \geq 0$? Explain your reasoning.

   b. Write the least positive number you can think of that is still a solution of the inequality $x > 0$. Explain your reasoning.

What Is Your Answer?

   4. IN YOUR OWN WORDS How can you use a number line to represent solutions of an inequality?

   5. Write an inequality. Graph all solutions of your inequality on a number line.

   6. Graph the inequalities $x > 9$ and $9 < x$ on different number lines. What do you notice?

Name _____ Date _____

7.5 Practice
For use after Lesson 7.5

Write the word sentence as an inequality.

1. A number n is at least 4.

2. A number x is less than 12.

Tell whether the given value is a solution of the inequality.

3. $4x \leq 20$; $x = 2$

4. $y + 5 > 8$; $y = 1$

Graph the inequality on a number line.

5. $x < 5$

6. $w \geq -\dfrac{1}{4}$

7. You buy tickets to a professional football game. You are allowed to buy at most 4 tickets. Write and graph an inequality to represent the number of tickets you are allowed to buy.

7.6 Solving Inequalities Using Addition or Subtraction
For use with Activity 7.6

Essential Question How can you use addition or subtraction to solve an inequality.

1 ACTIVITY: Writing an Inequality

Work with a partner. In 3 years, your friend will still not be old enough to vote.

   a. Which of the following represents your friend's situation?
What does x represent? Explain your reasoning.

| $x + 3 < 18$ | $x + 3 \leq 18$ | $x + 3 > 18$ | $x + 3 \geq 18$ |

   b. Graph the possible ages of your friend on a number line. Explain how you decided what to graph.

2 ACTIVITY: Writing an Inequality

Work with a partner. Baby manatees are about 4 feet long at birth. They grow to a maximum length of 13 feet.

   a. Which of the following can represent a baby manatee's growth?
What does x represent? Explain your reasoning.

| $x + 4 < 13$ | $x + 4 \leq 13$ | $x - 4 > 13$ | $x - 4 \geq 13$ |

   b. Graph the solution on a number line. Explain how you decided what to graph.

7.6 Solving Inequalities Using Addition or Subtraction (continued)

3 ACTIVITY: Solving Inequalities

Work with a partner. Complete the following steps for Activity 1. Then repeat the steps for Activity 2.

- Use your inequality from part (a). Replace the inequality symbol with an equal sign.

- Solve the equation.

- Replace the equal sign with the original inequality symbol.

- Graph this new inequality.

- Compare the graph with your graph in part (b). What do you notice?

7.6 **Solving Inequalities Using Addition or Subtraction** (continued)

4 **ACTIVITY:** The Triangle Inequality

Work with a partner. Draw different triangles whose sides have lengths 10 cm, 6 cm, and *x* cm.

x

10 cm

6 cm

a. Which of the following describes how *small x* can be? Explain your reasoning.

| $6 + x < 10$ | $6 + x \leq 10$ |

| $6 + x > 10$ | $6 + x \geq 10$ |

b. Which of the following describes how *large x* can be?

| $x - 6 < 10$ | $x - 6 \leq 10$ | $x - 6 > 10$ | $x - 6 \geq 10$ |

c. Graph the possible values of *x* on a number line.

What Is Your Answer?

5. IN YOUR OWN WORDS How can you use addition and subtraction to solve an inequality?

6. Describe a real-life situation that you can represent with an inequality. Write the inequality. Graph the solution on a number line.

Name _____ Date _____

Solve the inequality. Graph the solution.

1. $x + 6 \leq 15$

2. $y - 3 > 2$

3. $z + 1.5 \geq 2$

4. $p - \dfrac{1}{5} < \dfrac{7}{10}$

5. Your teacher gives you an assignment and says you have at most 2 weeks to complete the assignment. You are still working on the assignment after 5 days. Write and solve an inequality to represent how much more time you have to meet the requirement.

7.7 Solving Inequalities Using Multiplication or Division
For use with Activity 7.7

Essential Question How can you use multiplication or division to solve an inequality?

1 ACTIVITY: Writing an Inequality

Work with a partner. A store has a clearance rack of shirts that each cost the same amount. You buy 2 shirts and have money left after paying with a $20 bill.

a. Which of the following represents your purchase? What does x represent? Explain your reasoning.

$2x < 20$	$2x \leq 20$

$2x > 20$	$2x \geq 20$

b. Graph the possible values of x on a number line. Explain how you decided what to graph.

c. Can you buy a third shirt? Explain your reasoning.

2 ACTIVITY: Writing an Inequality

Work with a partner. One of your favorite stores is having a 75% off sale. You have $20. You want to buy a pair of jeans.

a. Which of the following represents your ability to buy the jeans with $20? What does x represent? Explain your reasoning.

$\frac{1}{4}x < 20$	$\frac{1}{4}x \leq 20$

$\frac{1}{4}x > 20$	$\frac{1}{4}x \geq 20$

7.7 Solving Inequalities Using Multiplication or Division (continued)

b. Graph the possible values of x on a number line. Explain how you decided what to graph.

c. Can you afford a pair of jeans that originally costs $100? Explain your reasoning.

---

3 ACTIVITY: Solving Inequalities

Work with a partner. Complete the following steps for Activity 1. Then repeat the steps for Activity 2.

- Use your inequality from part (a). Replace the inequality symbol with an equal sign.

- Solve the equation.

- Replace the equal sign with the original inequality symbol.

- Graph this new inequality.

- Compare the graph with your graph in part (b). What do you notice?

7.7 Solving Inequalities Using Multiplication or Division (continued)

4 **ACTIVITY:** Matching Inequalities

Work with a partner. Match the inequality with its graph.

a. $3x < 9$ **b.** $3x \leq 9$ **c.** $\dfrac{x}{2} \geq 1$

d. $6 < 2x$ **e.** $12 \leq 4x$ **f.** $\dfrac{x}{2} < 2$

A.

B.

C.

D.

E.

F.

What Is Your Answer?

5. **IN YOUR OWN WORDS** How can you use multiplication or division to solve an inequality?

7.7 Practice
For use after Lesson 7.7

Solve the inequality. Graph the solution.

1. $12q \geq 36$

2. $\dfrac{t}{4} > 6$

Graph the numbers that are solutions to both inequalities.

3. $6a \leq 42$ and $a + 4 > 7$

4. $d - 8 \leq 2$ and $9d < 81$

5. Each table in a banquet room seats 8 people. The room can seat no more than 360 people. Write and solve an inequality to represent the number of tables in the banquet room.

Name_____ Date_____

Identify the figure.

1.

2.

3.

4.

5. Identify the figure.

6 in.

6 in.   6 in.

6 in.

Chapter 8 Fair Game Review (continued)

Find the volume of the rectangular prism.

6.

7.

8.

9.

10.

11.

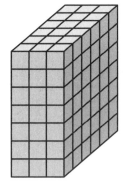

Name_____ Date_____

8.1 Three-Dimensional Figures
For use with Activity 8.1

Essential Question How can you draw three-dimensional figures?

Dot paper can help you draw three-dimensional figures, or *solids*.

Square Dot Paper

Face-on view

Isometric Dot Paper

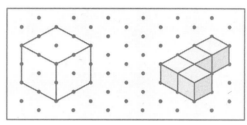

Corner view

1 ACTIVITY: Drawing Views of a Solid

Work with a partner. Draw the front, side, and top views of each stack of cubes. Then find the number of cubes in the stack.

a.

front	side	top

Number of cubes: ☐

b.

c.

d.

8.1 **Three-Dimensional Figures** (continued)

e.   **f.**   **g.**

2 **ACTIVITY:** Drawing Solids

Work with a partner.

a. Use isometric dot paper to draw three different solids that use the same number of cubes as the solid at the right.

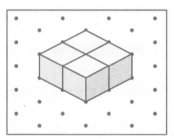

b. Use square dot paper to draw a different solid that uses the same number of *prisms* as the solid at the right.

8.1 Three-Dimensional Figures (continued)

3 ACTIVITY: Exploring Faces, Edges, and Vertices

Work with a partner. Use the solid shown.

a. Match each word to the figure. Then write a definition for each word.

face   edge   vertex

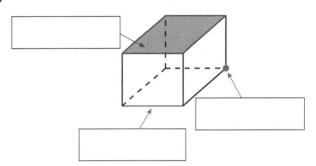

b. Identify the number of faces, edges, and vertices in a rectangular prism.

c. When using dot paper to draw a solid, what represents the vertices? How do you draw edges? How do you draw faces?

d. What do you think it means for lines or planes to be parallel or perpendicular in three dimensions? Use drawings to identify one pair of each of the following:

- parallel faces
- perpendicular faces

- parallel edges
- perpendicular edges

- edge parallel to a face
- edges perpendicular to a face

What Is Your Answer?

4. IN YOUR OWN WORDS How can you draw three-dimensional figures?

Name _____ Date _____

Draw the solid.

1. Pentagonal pyramid

2. Square prism

Draw the front, side, and top views of the solid.

3.

4.

5. Draw a solid with the following front, side, and top views.

Front   Side   Top

Name _____ Date _____

Essential Question How can you find the area of the entire surface of a prism?

1 ACTIVITY: Identifying Prisms

Work with a partner. Label one of the faces as a "base" and the other as a "lateral face." Use the shape of the base to identify the prism.

a.

_____ Prism

b.

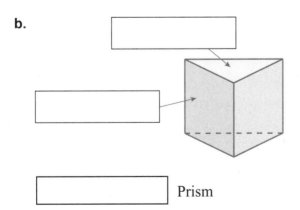

_____ Prism

8.2 **Surface Areas of Prisms** (continued)

2 **ACTIVITY:** Using Grid Paper to Construct a Prism

Work with a partner.

   a. Copy the figure shown below onto grid paper.*

   b. Cut out the figure and fold it to form a prism. What type of prism does it form?

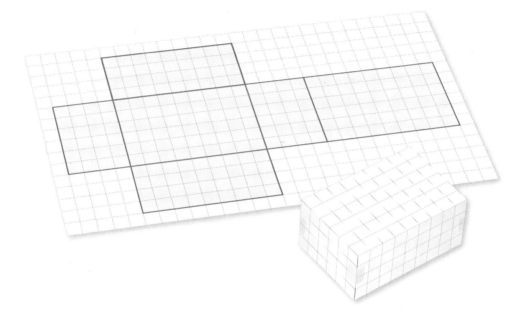

3 **ACTIVITY:** Finding the Area of the Entire Surface of a Prism

Work with a partner. Label each face in the two-dimensional representation of the prism as a "base" or a "lateral face." Then find the area of the entire surface of each prism.

   a.

*Cut-outs are available in the back of the Record and Practice Journal.

8.2 Surface Areas of Prisms (continued)

b.

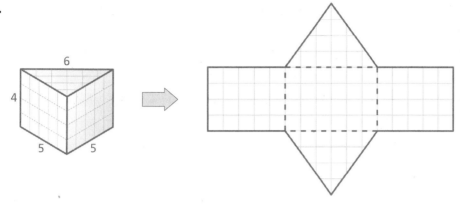

---

4 ACTIVITY: Drawing Two-Dimensional Representations of Prisms

Work with a partner. Draw a two-dimensional representation of each prism. Then find the area of the entire surface of each prism.

a.

b.

What Is Your Answer?

5. IN YOUR OWN WORDS How can you find the area of the entire surface of a prism?

8.2 Practice
For use after Lesson 8.2

Find the surface area of the prism.

1.

8 m

7 m

2 m

2.

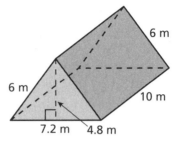

6 m

6 m

10 m

7.2 m   4.8 m

3.

6 cm

8 cm   12 cm

10 cm

4.

3 in.

10 in.

9 in.

5. You buy a ring box as a birthday gift that is in the shape of a triangular prism. What is the least amount of wrapping paper needed to wrap the box?

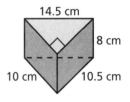

14.5 cm

8 cm

10 cm   10.5 cm

8.3 Surface Areas of Pyramids
For use with Activity 8.3

Essential Question   How can you use a net to find the surface area of a pyramid?

1 ACTIVITY: Identifying Pyramids

Work with a partner. Label one of the faces as a "base" and the other as a "lateral face." Use the shape of the base to identify the pyramid.

a.

b.

[   ] Pyramid   [   ] Pyramid

2 ACTIVITY: Using a Net

Work with a partner.

   a. Copy the net shown below onto grid paper.*

   b. Cut out the net and fold it to form a pyramid. What type of rectangle is the base? Use this shape to name the pyramid.

   c. Find the surface area of the pyramid.

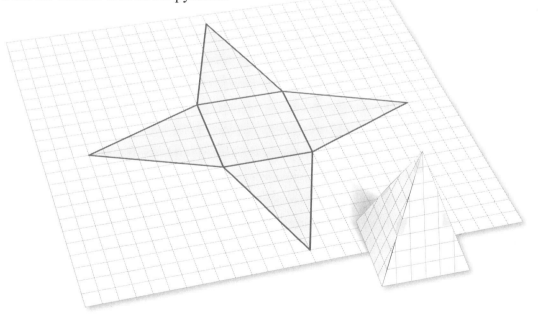

*Cut-outs are available in the back of the Record and Practice Journal.

8.3 Surface Areas of Pyramids (continued)

3 ACTIVITY: Estimating the Surface Area of a Triangular Pyramid

Work with a partner. Label each face in the net of the triangular pyramid as a "base" or a "lateral face." Then estimate the surface area of the pyramid.

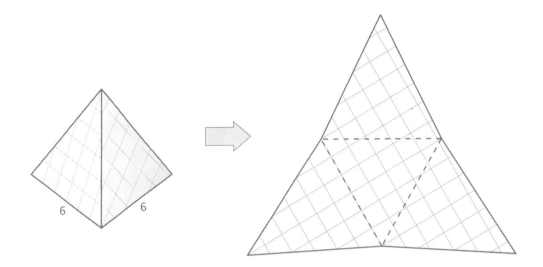

4 ACTIVITY: Finding the Surface Area of a Square Pyramid

Work with a partner. Draw a net for each square pyramid. Use the net to find the surface area of the pyramid.

a.

8.3 Surface Areas of Pyramids (continued)

b.

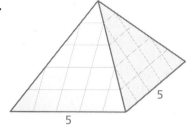

5

5

What Is Your Answer?

5. **IN YOUR OWN WORDS** How can you use a net to find the surface area of a pyramid?

6. **CONJECTURE** Make a conjecture about the lateral faces of a pyramid when the side lengths of the base have the same measure. Explain.

8.3 Practice
For use after Lesson 8.3

Find the surface area of the pyramid. The side lengths of the base are equal.

1.

7 cm

4 cm

2.

9 ft

6 ft

3.

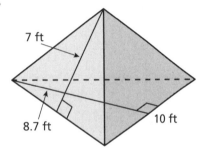

7 ft

8.7 ft

10 ft

4.

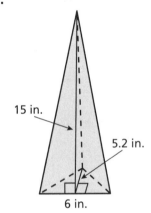

15 in.

5.2 in.

6 in.

5. A candle is shaped like a triangular pyramid. The side lengths of the base are equal. Find the surface area of the candle.

2 in.

2 in.

1.7 in.

8.4 Volumes of Rectangular Prisms
For use with Activity 8.4

Essential Question How can you find the volume of a rectangular prism with fractional edge lengths?

Recall that the **volume** of a three-dimensional figure is a measure of the amount of space that it occupies. Volume is measured in *cubic units*.

A *unit cube* is a cube with an edge length of 1 unit.

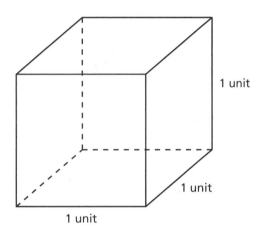

1 unit
1 unit
1 unit

---

1 ACTIVITY: Using a Unit Cube

Work with a partner. The parallel edges of the unit cube have been divided into 2, 3, and 4 equal parts to create smaller rectangular prisms that are identical.

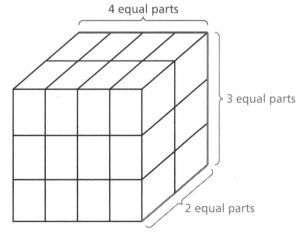

4 equal parts
3 equal parts
2 equal parts

8.4 **Volumes of Rectangular Prisms** (continued)

    a. Draw one of these identical prisms and label its dimensions.

    b. What fraction of the volume of the unit cube does one of these identical prisms represent? Use this value to find the volume of one of the identical prisms. Explain your reasoning.

2 **ACTIVITY:** Finding the Volume of a Rectangular Prism

Work with a partner.

    a. How many of the identical prisms in Activity 1(a) does it take to fill the rectangular prism below? Support your answer with a drawing.

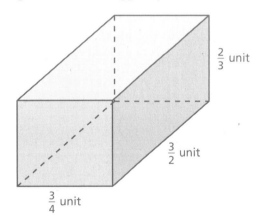

$\frac{2}{3}$ unit

$\frac{3}{2}$ unit

$\frac{3}{4}$ unit

    b. Use the volume of one of the identical prisms in Activity 1(a) to find the volume of the rectangular prism above. Explain your reasoning.

8.4 **Volumes of Rectangular Prisms** (continued)

3 **ACTIVITY:** Finding the Volumes of Rectangular Prisms

Work with a partner. Explain how you can use the procedure in Activities 1 and 2 to find the volume of each rectangular prism. Then find the volume of each prism.

a.

$\frac{3}{4}$ m

$\frac{1}{2}$ m $\frac{1}{2}$ m

b.

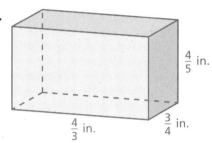

$\frac{4}{5}$ in.

$\frac{4}{3}$ in. $\frac{3}{4}$ in.

What Is Your Answer?

4. You have used the formulas $V = Bh$ and $V = \ell wh$ to find the volume V of a rectangular prism with whole number edge lengths. Do you think the formulas work for rectangular prisms with fractional edge lengths? Give examples with your answer.

5. **IN YOUR OWN WORDS** How can you find the volume of a rectangular prism with fractional edge lengths?

Name _____ Date _____

Practice
For use after Lesson 8.4

Find the volume of the prism.

1.

$\frac{2}{3}$ m

$\frac{1}{2}$ m

$\frac{1}{3}$ m

2.

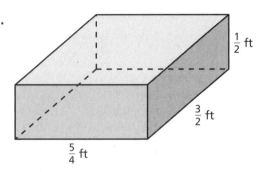

$\frac{1}{2}$ ft

$\frac{3}{2}$ ft

$\frac{5}{4}$ ft

Write and solve an equation to find the missing dimension of the prism.

3. Volume = 18,000 in.^{3}

30 in.

x

30 in.

4. Volume = 55 in.^{3}

11 in.

ℓ

0.5 in.

5. You are mailing a birthday present to a friend. You have a box that has a length of $2\frac{1}{2}$ feet, a height of 2 feet, and a width of $1\frac{1}{2}$ feet. The present has a volume of 3 cubic feet. What is the volume of the empty space in the box?

Name_____ Date_____

Chapter 9 Fair Game Review

Use a number line to order the numbers from least to greatest.

1. 1.5, 4.5, 5, 2.5, 1, 3

2. 6, 3.5, 4, 5.5, 7.5, 4.5

3. 5.25, 6, 3.5, 5, 6.25, 4.25

4. 4.75, 6.5, 7, 7.75, 5.5, 3

5. 3.25, 5.75, 4.5, 3.75, 4.25, 6.5

6. 3.75, 1.5, 4.75, 1.25, 2.25, 3.5

Chapter 9 **Fair Game Review** (continued)

In Exercises 7–9, use the double bar graph that shows the sales of a clothing store over two days.

7. How much more did the store earn selling shirts on Day 1 than on Day 2?

8. Which item had the largest change in sales?

9. Which item had the highest sales total for the two days?

Name_____ Date_____

9.1 Introduction to Statistics
For use with Activity 9.1

Essential Question How can you tell whether a question is a statistical question?

1 ACTIVITY: Using Data to Answer Questions

Work with a partner.

a. Find your pulse by counting the number of beats in 10 seconds. Have your partner keep track of the time. Write a rate to describe your result.

b. Complete the ratio table. What is your heart rate in beats per minute?

Time (seconds)	10	30	60
Number of beats			

c. Collect the recorded heart rates (in beats per minute) of the students in your class, including yourself. Compare the heart rates.

d. **MODELING** Make a *line plot* of your data. Then answer the following questions:

- How many values are in your data set?

- Do the heart rates *cluster* around a particular value or values?

- Are there any *peaks* or *gaps* in the data?

9.1 Introduction to Statistics (continued)

- Are there any unusual heart rates that are far removed from the other values?

e. **REASONING** How would you answer the following question by using only one value? Explain your reasoning.

"What is the heart rate of sixth grade students?"

f. **REASONING** Read and compare the following questions. How did you answer each question? Could the answer be the same for both questions? Explain.

- *What is your heart rate?*
- *What is the heart rate of sixth grade students?*

2 ACTIVITY: Identifying Types of Questions

Work with a partner.

a. Answer each question below on your own. Then compare your answers with your partner's answers. For which questions should your answers be the same? For which questions might your answers be different?

1. What is your shoe size?

2. How many states are in the United States?

3. How many brothers and sisters do you have?

4. How many U.S. presidents have been in office?

5. What is your favorite type of movie?

6. How tall are you?

9.1 **Introduction to Statistics** (continued)

b. CONJECTURE Some of the questions on the previous page are considered *statistical* questions. Which ones do you think they are? Why?

3 **ACTIVITY:** Analyzing a Question in a Survey

Work with a partner. A student asks the following question in a survey:

"Do you prefer salty potato chips or healthy granola bars to be sold in the school's vending machines?"

a. Do you think this is a fair question to ask in a survey? Explain.

b. LOGIC Identify the words in the question that may influence someone's response. Then explain how you can reword the question.

c. How might the results of the survey differ when the student asks the original question and your reworded question in part (b)?

What Is Your Answer?

4. REASONING What do you think "statistics" means?

5. IN YOUR OWN WORDS How can you tell whether a question is a statistical question? Give examples to support your explanation.

6. Find the least and the greatest heart rates in your class. How can you use these two values to answer the question in Activity 1(e)?

7. Create a one-question survey. Explain why your question is a statistical question. Then conduct your survey and organize your results in a line plot. Make three observations about your data set.

Name _____ Date _____

Tell whether the question is a statistical question. Explain.

1. How many songs are on your MP3 player?

2. How many feet are in 1 yard?

3. What is the name of your favorite movie?

4. How many sixth graders attend your school?

5. The dot plot shows the numbers of text messages received by students over the weekend.

   a. How many students are represented?

   b. How can you collect this data? What are the units?

   c. Write a statistical question that you can answer using the dot plot. Then answer the question.

9.2 Mean

For use with Activity 9.2

Essential Question How can you find an average value of a data set?

1 **ACTIVITY:** Finding a Balance Point

Work with a partner. Discuss the distribution of the data. Where on the number line do you think the data set is *balanced*? Is this a good representation of the average? Explain.

    a. number of quarters brought to a batting cage

    b. annual income of recent graduates (in thousands of dollars)

    c. hybrid fuel economy (miles per gallon)

9.2 **Mean** (continued)

2 **ACTIVITY:** Finding a Fair Share

Work with a partner. It costs \$0.25 to hit 12 baseballs in a batting cage. The table shows the numbers of quarters six friends bring to the batting cage. They want to group the quarters so that everyone has the same amount.

Quarters					
John	Lisa	Miguel	Matt	Cheryl	Jean
6	3	4	5	2	4

Use counters to represent each number in the table. How can you use the counters to determine how many times each friend can use the batting cage? Explain how this procedure results in a "fair share."

3 **ACTIVITY:** Finding an Average

Work with a partner. Use the information in Activity 2.

a. What is the total number of quarters the group of friends brought to the batting cage?

b. **REASONING** How can you use math to find the average number of quarters that each friend brought to the batting cage? Find the average number of quarters. Why do you think this average represents a fair share?

4 **ACTIVITY:** Answering a Statistical Question

Work with a partner. The table shows the numbers of quarters several people bring to a batting cage. You want to answer the question:

"How many quarters do people bring to the batting cage?"

a. Explain why this question is a statistical question.

Quarters			
6	8	8	12
8	12	8	4
8	6	6	10
7	10	7	8

Name_____ Date _____

b. **MODELING** Make a dot plot of the data. Use the distribution of the data to answer the question. Explain your reasoning.

c. **REASONING** Use an average to answer the question. Explain your reasoning.

What Is Your Answer?

5. **IN YOUR OWN WORDS** How can you find an average value of a data set?

6. Give two real-life examples of averages.

7. Explain what it means to say the average of a data set is the point on a number line where the data set is balanced.

8. There are 5 students in the cartoon. Four of the students are 66 inches tall. One is 96 inches tall.

a. How do you think the students decided their average height is 6 feet?

"Yup, the average height in our class is 6 feet."

b. Does a height of 6 feet seem like a good representation of the average height of the 5 students? Explain why or why not.

9.2 Practice
For use after Lesson 9.2

Find the mean of the data.

1. Emails sent in the last 4 hours:

   2, 5, 4, 5

2. Magazine subscriptions sold this week:

   3, 6, 7, 6, 7, 9, 11

3.

Books Brought Home	
Monday	|
Tuesday	|||
Wednesday	||||
Thursday	||||
Friday	|

4. The table shows the number of points scored by your team in each quarter of a football game. What is the mean number of points scored in a quarter?

Quarter	1	2	3	4
Points	3	14	10	0

9.3 Measures of Center
For use with Activity 9.3

Essential Question In what other ways can you describe an average of a data set?

---

1 **ACTIVITY:** Finding a Median

Work with a partner.

  a. Write the total number of letters in the first and last names of 19 celebrities, historical figures, or people you know. Organize your data in a table. One person is already listed for you.

Person	Number of letters in first and last name
Abraham Lincoln	14

  b. Order the values in your data set from least to greatest. Then write the data on a strip of grid paper with 19 boxes.

  c. Place a finger on the square at each end of the strip. Move your fingers toward the center of the ordered data set until your fingers touch. On what value do your fingers touch?

  d. Now take your strip of grid paper and fold it in half. On what number is the crease? What do you notice? This value is called the *median*. How would you describe to another student what the median of a data set represents?

  e. How many values are greater than the median? How many are less than the median?

9.3 **Measures of Center** (continued)

    f. Why do you think the median is considered an average of a data set?

2 **ACTIVITY:** Adding a Value to a Data Set

Work with a partner.

    a. How many total letters are in your first name and last name? Add this value to the ordered data set in Activity 1. How many values are now in your data set?

    b. Write the ordered data, including your new value from part (a), on a strip of grid paper.

    c. Repeat parts (c) and (d) from Activity 1. Explain your findings. How do you think you can find the median of this data set?

    d. Compare the medians in Activities 1 and 2. Then answer the following questions. Explain your reasoning.

       • Do you think the median always has to be a value in the data set?

       • Do you think the median always has to be a whole number?

3 **ACTIVITY:** Finding a Mode

Work with a partner.

    a. Make a dot plot for the data set in Activity 2. Describe the distribution of the data.

9.3 **Measures of Center** (continued)

b. Which value occurs most often in the data set? This value is called the *mode*.

c. Do you think a data set can have no mode or more than one mode? Explain.

d. Do you think the mode always has to be a value in the data set? Explain.

e. Why do you think the mode is considered an average of a data set?

What Is Your Answer?

4. **IN YOUR OWN WORDS** In what other ways can you describe an average of a data set?

5. Find the mean of your data set in Activity 2. Then compare the mean, median, and mode. Is there one measure that you think best represents your data set? Explain your reasoning.

Name _____ Date _____

Find the median and mode(s) of the data.

1. 3, 2, 3, 6, 7, 5, 9

2. 17, 21, 30, 17, 28, 21

Find the mean, median, and mode(s) of the data with and without the outlier. Describe the effect of the outlier on the measures of center.

3. 4, 15, 6, 12, 68, 12

4. 0, 54, 62, 64, 55, 55, 54, 62

5. The data show your strokes for 18 holes of miniature golf.

4, 5, 3, 3, 1, 2, 3, 2, 4, 8, 2, 4, 4, 5, 2, 3, 6, 2

Find the mean, median, and mode(s) of the data. Which measure best represents the data? Explain your reasoning.

Name_____   Date _____

9.4   Measures of Variation
For use with Activity 9.4

Essential Question   How can you describe the spread of a data set?

---

1   ACTIVITY: Interpreting Statements

Work with a partner. There are 24 students in your class. Your teacher makes the following statements:

- *"The exam scores range from 75% to 96%."*

- *"Most of the students received high scores."*

a. What do you think the first statement means? Explain.

b. In the first statement, is your teacher describing the center of the data set? If not, what do you think your teacher is describing?

c. What do you think the scores are for most of the students in the class? Explain your reasoning.

d. Use your teacher's statements to make a dot plot that can represent the distribution of the exam scores of the class.

---

2   ACTIVITY: Grouping Data

Work with a partner. The numbers of U.S. states visited by each student in a sixth grade class are shown.

a. Between what values do the data range?

b. Write the ordered data values on a strip of grid paper and fold it to find the median. How many values are greater than the median? How many are less than the median?

Number of States Visited			
1	7	5	2
11	6	3	20
4	18	1	6
2	7	1	8
10	2	12	5
	3	21	

9.4   **Measures of Variation** (continued)

   c. **REPEATED REASONING** Fold the strip in half again. On what values are the two new creases? What do you think these values represent?

   d. Into how many parts did you divide the data set? How many data values are in each part?

   e. Graph the median and the values you found in parts (a) and (c) on a number line. Are the distances the same between these points?

   f. How can you use these values to describe the spread of the data?

3   **ACTIVITY:** Adding a Value to a Data Set

Work with a partner. A new student joins the class in Activity 2. The new student has visited 41 states.

   a. Add this value to the ordered data set in Activity 2. Does your answer to part (a) change? Explain.

   b. How does the distribution of the data change when this value is added? Explain your reasoning.

   c. How does adding this value affect the values on your number line in part (e) of Activity 2?

9.4 **Measures of Variation** (continued)

4 **ACTIVITY:** Analyzing Data Sets

Work with a partner. Identify the data set that is the least spread out and the data set that is the most spread out. Explain your reasoning.

a.

b.

c.

d.

What Is Your Answer?

5. **IN YOUR OWN WORDS** How can you describe the spread of a data set?

6. Make a dot plot of the data set in Activity 2. Describe any similarities between the dot plot and the number line in part (e).

Name _____ Date _____

Find the range of the data.

1. 3, 1, 6, 10, 12, 2

2. 9, 13, 8, 7, 14, 16

3. 11, 17, 21, 23, 19, 16

4. 29, 37, 27, 28, 31, 33

Find the median, first quartile, third quartile, and interquartile range of the data.

5. 20, 12, 14, 22, 18, 21, 24, 15

6. 37, 28, 30, 40, 31, 27, 33, 42, 43

7. The table shows the number of song downloads a group of friends made on Saturday.

Numbers of Songs			
12	6	9	14
15	16	8	10

a. Find and interpret the range of the numbers of song downloads.

b. Find and interpret the interquartile range of the numbers of song downloads.

9.5 Mean Absolute Deviation
For use with Activity 9.5

Essential Question How can you use the distance between each data value and the mean of a data set to measure the spread of a data set?

1 **ACTIVITY:** Finding Distances from the Mean

Work with a partner. The table shows the exam scores of 14 students in your class.

   a. What is the mean exam score?

Exam Scores			
Ben	89	Mike	95
Emma	86	Hong	96
Jeremy	80	Rob	92
Pete	80	Amy	90
Ryan	96	Sue	76
Dan	94	Kim	84
Lucy	89	Heather	85

   b. Make a dot plot of the data. Place an "X" on the number line to represent the mean.

   c. Is the number of exam scores that are greater than the mean equal to the number of exam scores that are less than the mean? Explain.

   d. Which exam score *deviates* the most from the mean? Which exam score *deviates* the least from the mean? Explain how you found your answers.

   e. Overall, do you think the exam scores are *close to* the mean or *far away* from the mean? Explain your reasoning.

9.5 **Mean Absolute Deviation** (continued)

2 ACTIVITY: Using Distances From the Mean

Work with a partner. Use the information in Activity 1.

a. Complete the table below. Add rows if needed. Be sure to find the sum of the values in the last column of the table.

Student with Score *Less Than* the Mean	Exam Score	Distance from the Mean
	Sum:	

b. Complete the table for students with scores greater than the mean.

Student with Score *Greater Than* the Mean	Exam Score	Distance from the Mean
	Sum:	

c. **LOGIC** What do you notice about the sums you found in your tables? Why do you think this happens?

9.5 **Mean Absolute Deviation** (continued)

3 **ACTIVITY:** Interpreting Distances From the Mean

Work with a partner.

a. **LOGIC** Add the sums you found in your tables in Activity 2. Divide that amount by the total number of students. Round your result to the nearest tenth.

In your own words, what do you think this value represents?

b. **REASONING** In a data set, what do you think it means when the value you found in part (a) is close to 0? Explain.

What Is Your Answer?

4. **IN YOUR OWN WORDS** How can you use the distances between each data value and the mean of a data set to measure the spread of a data set?

5. **REASONING** Find the range and the interquartile range of the data set in Activity 1. What do you think it means when these values are close to 0? Explain.

Name _____ Date _____

Find and interpret the mean absolute deviation of the data. Round your answer to the nearest tenth, if necessary.

1.

Price of Textbooks (dollars)				
78	99	90	80	55
56	102	88	60	42

2.

Numbers of Songs on an Album				
10	13	7	12	9
8	12	10	11	13

3.

Height of Plants (inches)				
1	7	10	5	3
3	6	12	9	4

4.

Numbers of Applications on a Smart Phone				
30	46	25	45	18
25	15	32	40	24

5. The data set shows the admission prices at several amusement parks.

   $16, $25, $12, $20, $10, $25

   Find and interpret the range, interquartile range, and mean absolute deviation of the data.

Name_____ Date_____

The bar graph shows the favorite types of salad dressings of the students in a class.

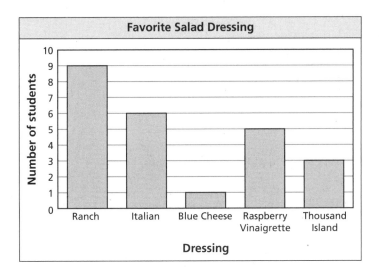

1. What salad dressing was chosen the most?

2. How many students said Raspberry Vinaigrette or Thousand Island is their favorite salad dressing?

3. How many students did *not* choose Italian as their favorite salad dressing?

4. How many students are in the class?

Chapter 10 **Fair Game Review** (continued)

The circle graph shows the results from a class survey on favorite juice. There are 30 students in the class.

Favorite Juice

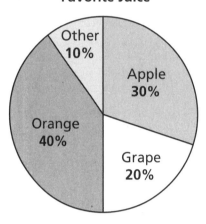

5. How many students said their favorite juice is apple?

6. How many students said their favorite juice is orange?

7. How many students said their favorite juice is grape?

10.1 Stem-and-Leaf Plots
For use with Activity 10.1

Essential Question How can you place values to represent data graphically?

---

1 ACTIVITY: Making a Data Display

Work with a partner. The list below gives the ages of these women when they became first ladies of the United States.

THE WHITE HOUSE
WASHINGTON

Frances Cleveland – 21   Mamie Eisenhower – 56
Caroline Harrison – 56   Jacqueline Kennedy – 31
Ida McKinley – 49   Claudia Johnson – 50
Edith Roosevelt – 40   Patricia Nixon – 56
Helen Taft – 48   Elizabeth Ford – 56
Ellen Wilson – 52   Rosalynn Carter – 49
Florence Harding – 60   Nancy Reagan – 59
Grace Coolidge – 44   Barbara Bush – 63
Lou Hoover – 54   Hillary Clinton – 45
Eleanor Roosevelt – 48   Laura Bush – 54
Elizabeth Truman – 60   Michelle Obama – 45

a. The incomplete data display shows the ages of the first ladies in the left column of the list above.

Ages of First Ladies

2	1
3	
4	0 4 8 8 9
5	2 4 6
6	0 0

What do the numbers to the left of the line represent? What do the numbers to the right of the line represent?

10.1 Stem-and-Leaf Plots (continued)

b. This data display is called a *stem-and-leaf plot*. What numbers do you think represent the *stems*? *leaves*? Explain your reasoning.

c. Complete the stem-and-leaf plot on the previous page using the remaining ages in the right column. Order the numbers to the right of the line in numerical order.

d. Write a question about the ages of first ladies that would be easier to answer using a stem-and-leaf plot than a dot plot.

---

2 ACTIVITY: Making a Back-to-Back Stem-and-Leaf Plot

Work with a partner. The table below shows the ages of presidents of the United States from 1885 to 2009 on their first inauguration day.

Ages of Presidents										
47	55	54	42	51	56	55	51	54	51	60
62	43	55	56	61	52	69	64	46	54	47

a. On your stem-and-leaf plot from Activity 1(c), draw a vertical line to the left of the display. Represent the ages of the presidents by including numbers to the left of the line.

b. Find the median ages of both the first ladies and presidents of the United States.

c. Compare the distribution of each data set.

10.1 **Stem-and-Leaf Plots** (continued)

3 **ACTIVITY:** Conducting an Experiment

Work with a partner. Use two number cubes to conduct the following experiment.

- Toss the cubes and find the product of the resulting numbers.

- Repeat this process 30 times. Record your results.

Toss	1	2	3	4	5	6	7	8	9	10
Product										
Toss	11	12	13	14	15	16	17	18	19	20
Product										
Toss	21	22	23	24	25	26	27	28	29	30
Product										

a. Use a stem-and-leaf plot to organize your results.

b. Describe the distribution of the data.

What Is Your Answer?

4. IN YOUR OWN WORDS How can you use place values to represent data graphically?

5. How can you display data in a stem-and-leaf plot whose values range from 82 through 129?

Name _____ Date _____

10.1 Practice
For use after Lesson 10.1

Make a stem-and-leaf plot of the data.

1.

Class Sizes			
12	10	21	28
9	16	19	16
25	32	14	21

2.

Minutes Spent on Homework			
75	82	91	68
92	86	79	76
75	81	88	60

3. The number of text messages from eight phones are 8, 11, 14, 22, 5, 15, 7, and 20. Make a stem-and-leaf plot of the data. Describe the distribution of the data.

4. The number of minutes seven members spent at band practice are 57, 49, 55, 62, 78, 72, and 75. Make a stem-and-leaf plot of the data. Describe the distribution of the data.

5. The stem-and-leaf plot shows the numbers of miles students travel to get to school.

a. How many students travel more than 15 miles?

b. Find the mean, median, mode, range, and interquartile range of the data.

Stem	Leaf
0	5 7
1	2 4 8
2	0 1 5 7
3	3

Key: 1 | 4 = 14 miles

10.2 Histograms
For use with Activity 10.2

Essential Question How can you use intervals, tables, and graphs to help organize data?

---

1 ACTIVITY: Conducting an Experiment

Work with a partner.

a. Roll a number cube 20 times. Record your results in a tally chart.

	1	2	3	4	5	6
Tally						

Key: | = 1   ⵚ = 5

b. Make a bar graph of the totals.

c. Go to the board and enter your totals in the class tally chart.

d. Make a second bar graph showing the class totals. Compare and contrast the two bar graphs.

10.2 **Histograms** (continued)

2 **ACTIVITY:** Using Intervals to Organize Data

Work with a partner. You are judging a paper airplane contest. A contestant flies a paper airplane 20 times. You record the following distances:

20.5 ft, 24.5 ft, 18.5 ft, 19.5 ft, 21.0 ft, 14.0 ft, 12.5 ft, 20.5 ft, 17.5 ft, 24.5 ft,
19.5 ft, 17.0 ft, 18.5 ft, 12.0 ft, 21.5 ft, 23.0 ft, 13.5 ft, 19.0 ft, 22.5 ft, 19.0 ft

  a. Complete the tally chart and the bar graph of the distances.

Interval	Tally	Total
10.0–12.9		
13.0–15.9		
16.0–18.9		
19.0–21.9		
22.0–24.9		

  b. Make a different tally chart and bar graph of the distances. Use the following intervals:

     10.0–11.9, 12.0–13.9, 14.0–15.9, 16.0–17.9,
     18.0–19.9, 20.0–21.9, 22.0–23.9, 24.0–25.9

  c. Which graph do you think represents the distances better? Explain.

10.2 **Histograms** (continued)

3 **ACTIVITY:** Developing an Experiment

Work with a partner.

a. Make the airplane shown in your textbook from a single sheet of $8\frac{1}{2}$-by-11 inch paper. Then design and make your own paper airplane.

b. **PRECISION** Fly each airplane 20 times. Keep track of the distance flown each time.

Flight	1	2	3	4	5	6	7	8	9	10
Plane A										
Plane B										
Flight	11	12	13	14	15	16	17	18	19	20
Plane A										
Plane B										

c. **MODELING** Organize the results of the flights using frequency tables and graphs. Which airplane flies farther? Explain your reasoning.

What Is Your Answer?

4. **IN YOUR OWN WORDS** How can you use intervals, tables, and graphs to organize data?

5. What intervals could you use in a graph that displays data whose values range from 40 through 59?

Name _____ Date _____

10.2 Practice
For use after Lesson 10.2

Display the data in a histogram.

1.

Birthdays	
Months	**Frequency**
Jan–Mar	15
Apr–June	9
Jul–Sept	6
Oct–Dec	12

2.

Goals Scored	
Goals	**Frequency**
0–2	6
3–5	8
6–8	2
9–11	1

3.

Height Jumped	
Inches	**Frequency**
0–11	7
12–23	10
24–35	5
36–47	2

4.

Money Spent	
Dollars	**Frequency**
0–19	3
20–39	8
40–59	8
60–79	15

5. The histogram shows the times students ran the mile in gym class.

  a. Which interval contains the fewest data values?

  b. How many students are in the class?

  c. What percent of students ran the mile in 12 minutes or less?

Name_____ Date_____

10.3 Shapes of Distributions

For use with Activity 10.3

Essential Question How can you describe the shape of the distribution of a data set?

---

1 ACTIVITY: Describing the Shape of a Distribution

Work with a partner. The lists at the right show the last four digits of a set of phone numbers in a phone book.

  a. Create a list that represents the last digit of each phone number shown. Make a dot plot of the data.

  b. In your own words, how would you describe the shape of the distribution? What single word do you think you can use to identify this type of distribution? Explain your reasoning.

```
-7253      -8678
-7290      -2063
-7200      -2911
-1192      -2103
-1142      -4328
           -7826
-3500      -7957
-2531      -7246
-2079      -2119
-5897      -7845
-5341      -1109
-1392      -9154
-5406
-7875
-7335
-0494
-9018
-2184
-2367
```

---

2 ACTIVITY: Describing the Shape of a Distribution

Work with a partner. The lists at the right show the first three digits of a set of phone numbers in a phone book.

  a. Create a list that shows the first digit of each phone number shown. Make a dot plot of the data.

```
538-      664-
438-      664-
664-      538-
761-      855-
868-      664-
          538-
735-      654-
694-      654-
599-      725-
725-      538-
556-      799-
555-      764-
456-
736-
664-
576-
664-
664-
725-
```

10.3 **Shapes of Distributions** (continued)

b. In your own words, how would you describe the shape of the distribution? What single word do you think you can use to identify this type of distribution? Explain your reasoning.

c. In your dot plot, draw a vertical line through the middle of the data set. What do you notice?

d. Repeat part (c) for the dot plot you constructed in Activity 1. What do you notice? Compare the distributions from Activities 1 and 2.

3 **ACTIVITY:** Describing the Shape of a Distribution

Work with a partner. The table shows the ages of cellular phones owned by a group of students.

Ages of Cellular Phones (years)				
0	1	0	6	4
2	3	5	1	1
0	1	2	3	1
0	0	1	1	1
7	1	4	2	2
0	2	0	1	2

a. Make a dot plot of the data.

b. In your own words, how would you describe the shape of the distribution? Compare it to the distributions in Activities 1 and 2.

10.3 Shapes of Distributions (continued)

   c. Why do you think this type of distribution is called a *skewed distribution*?

---

4   **ACTIVITY:** Finding Measures of Center

Work with a partner.

   a. Find the mean and median of the data sets in Activities 1–3.

   b. What do you notice about the means and medians of the data sets and the shapes of the distributions? Explain.

   c. Which measure of center do you think best describes the data set in Activity 2? in Activity 3? Explain your reasoning.

   d. Using your answers to part (c), decide which measure of variation you think best describes the data set in Activity 2. Which measure of variation do you think best describes the data set in Activity 3? Explain your reasoning.

What Is Your Answer?

   5. IN YOUR OWN WORDS How can you describe the shape of the distribution of a data set?

   6. Name two other ways you can describe the distribution of a data set.

Name _____ Date _____

Describe the shape of each distribution.

1.

Gift Bags

2.

Sidewalks

3.

Concert Length

4.
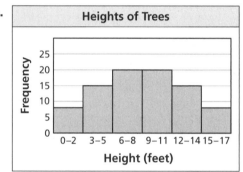
Heights of Trees

5. The frequency table shows the number of months each person has been a member of a golf league. Display the data in a histogram. Describe the shape of the distribution.

Months as a Member	0–4	5–9	10–14	15–19	20–24	25–29	30–34
Frequency	4	6	8	10	12	12	8

Name_____ Date_____

Choose the most appropriate measures to describe the center and the variation. Find the measures you chose.

1.   **Prices of Shirts**

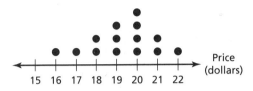

2.   **Weekly Triathlon Training Times**

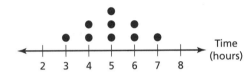

3.   **Number of Game Downloads**

4.   **Plant Heights**

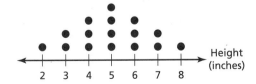

Extension 10.3 **Practice** (continued)

5. The frequency table shows the number of pages of the science reports written by each student in the sixth-grade class.

Number of pages	Frequency
2–4	5
5–7	16
8–10	12
11–13	7
14–16	4
17–19	0

   a. Display the data in a histogram.

   b. What are the most appropriate measures to describe the center and the variation?

Name_____ Date _____

10.4 Box-and-Whisker Plots
For use with Activity 10.4

Essential Question How can you use quartiles to represent data graphically?

1 ACTIVITY: Drawing a Box-and-Whisker Plot

Work with a partner.

The numbers of pairs of footwear owned by each student in a sixth grade class are shown.

A box-and-whisker plot uses a number line to represent the data visually.

Numbers of Pairs of Footwear			
2	5	12	3
7	2	4	6
14	10	6	28
5	3	2	4
9	25	4	10
8	15	5	8

a. Order the data set from least to greatest. Then write the data on a strip of grid paper with 24 boxes.

b. Use the strip of grid paper to find the median, the first quartile, and the third quartile. Identify the least value and the greatest value in the data set.

c. Graph the five numbers that you found in part (b) on the number line below.

10.4 **Box-and-Whisker Plots** (continued)

d. The data display shown below is called a *box-and-whisker plot*. Fill in the missing labels and numbers. Explain how the box-and-whisker plot uses quartiles to represent the data.

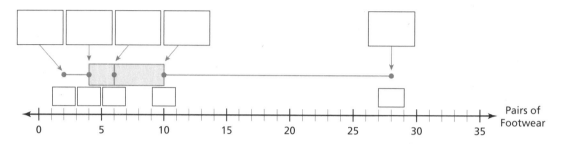

e. Using only the box-and-whisker plot, which measure(s) of center can you find for the data set? Which measure(s) of variation can you find for the data set? Explain your reasoning.

f. Why do you think this type of data display is called a box-and-whisker plot? Explain.

2 **ACTIVITY:** Conducting a Survey

Have your class conduct a survey. Each student will write on the chalkboard the number of pairs of footwear that he or she owns.

Now, work with a partner to draw a box-and-whisker plot of the data.

Name_____ Date _____

3 **ACTIVITY:** Reading a Box-and-Whisker Plot

Work with a partner. The box-and-whisker plots show the test score distributions of two sixth grade achievement tests. The same group of students took both tests. The students took one test in the fall and the other in the spring.

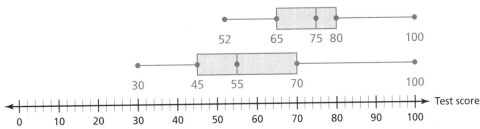

a. Compare and contrast the test results.

b. Decide which box-and-whisker plot represents the results of which test. How did you make your decision?

What Is Your Answer?

4. IN YOUR OWN WORDS How can you use quartiles to represent data graphically?

5. Describe who might be interested in test score distributions like those shown in Activity 3. Explain why it is important for such people to know test score distributions.

10.4 Practice
For use after Lesson 10.4

Make a box-and-whisker plot for the data.

1. Test scores: 63, 57, 52, 62, 60, 59, 55, 62, 61, 56

2. Pairs of sunglasses: 1, 3, 1, 2, 4, 5, 3, 6, 7

3. Miles: 18, 12, 25, 22, 15, 30, 28, 21, 27, 22, 16, 23

4. Numbers of photos: 32, 28, 36, 38, 40, 26, 29, 37

5. The numbers of times you woke up in the middle of the night over the past week are 3, 0, 2, 1, 3, 4, and 1. Make a box-and-whisker plot for the data.

Chapter 11 Fair Game Review

Simplify the expression. Explain each step.

1. $2 + (5 + y)$

2. $(c + 1) + 9$

3. $(2.3 + n) + 1.4$

4. $7 + (d + 5)$

5. $10(7t)$

6. $8(4k)$

Chapter 11 **Fair Game Review** (continued)

7. $13 \bullet 0 \bullet p$

8. $7 \bullet z \bullet 0$

9. $2.5 \bullet w \bullet 1$

10. $1 \bullet x \bullet 19$

11. $(t + 3) + 0$

12. $0 + (g + 4)$

11.1 Integers and Absolute Value
For use with Activity 11.1

Essential Question How can you use integers to represent the velocity and the speed of an object?

On these three pages, you will investigate vertical motion (up or down).

- Speed tells how fast an object is moving, but it does not tell the direction.

- Velocity tells how fast an object is moving, and it also tells the direction.

  When velocity is positive, the object is moving up.

  When velocity if negative, the object is moving down.

1 ACTIVITY: Falling Parachute

Work with a partner. You are gliding to the ground wearing a parachute. The table shows your height above the ground at different times.

Time (seconds)	0	1	2	3
Height (feet)	90	75	60	45

a. Describe the pattern in the table. How many feet do you move each second? After how many seconds will you land on the ground?

b. What integer represents your speed? Give the units.

c. Do you think your velocity should be represented by a positive or negative integer? Explain your reasoning.

d. What integer represents your velocity? Give the units.

Name _____ Date _____

Integers and Absolute Value (continued)

2 ACTIVITY: Rising Balloons

Work with a partner. You release a group of balloons. The table shows the height of the balloons above the ground at different times.

Time (seconds)	0	1	2	3
Height (feet)	8	12	16	20

a. Describe the pattern in the table. How many feet do the balloons move each second? After how many seconds will the balloons be at a height of 40 feet?

b. What integer represents the speed of the balloons? Give the units.

c. Do you think the velocity of the balloons should be represented by a positive or negative integer? Explain your reasoning.

d. What integer represents the velocity of the balloons? Give the units.

3 ACTIVITY: Firework Parachute

Work with a partner. The table shows the height of a firework's parachute above the ground at different times.

Time (seconds)	Height (feet)
0	480
1	360
2	240
3	120
4	0

a. Describe the pattern in the table. How many feet does the parachute move each second?

b. What integer represents the speed of the parachute? What integer represents the velocity? How are these integers similar in their relation to 0 on a number line?

242 **Big Ideas Math Advanced 1**
Record and Practice Journal

Copyright © Big Ideas Learning, LLC
All rights reserved.

11.1   **Integers and Absolute Value** (continued)

Inductive Reasoning

4. Complete the table.

Velocity (feet per second)	−14	20	−2	0	25	−15
Speed (feet per second)						

5. Find two different velocities for which the speed is 16 feet per second.

6. Which number is greater: −4 or 3? Use a number line to explain your reasoning.

7. One object has a velocity of −4 feet per second. Another object has a velocity of 3 feet per second. Which object has the greater speed? Explain your answer.

8. **IN YOUR OWN WORDS** How can you use integers to represent the velocity and the speed of an object?

What Is Your Answer?

9. **LOGIC** In this lesson, you will study **absolute value**. Here are some examples:

   Absolute value of $|-16| = 16$ Absolute value of $|16| = 16$

   Absolute value of $|0| = 0$ Absolute value of $|-2| = 2$

   Which of the following is a true statement? Explain your reasoning.

   | $|\text{velocity}| = \text{speed}$ | | $|\text{speed}| = \text{velocity}$ |

Name _____ Date _____

Find the absolute value.

1. $|-1|$

2. $|-14|$

3. $|0|$

4. $|6|$

Complete the statement using <, >, or =.

5. $6 \underline{\quad} |-2|$

6. $-7 \underline{\quad} |-8|$

7. $|-9| \underline{\quad} 5$

8. $|-2| \underline{\quad} 2$

Order the values from least to greatest.

9. $4, |7|, -1, |-3|, -4$

10. $|2|, -3, |-5|, -1, 6$

11. You download 12 new songs to your MP3 player. Then you delete 5 old songs. Write each amount as an integer.

Name_____ Date_____

11.2 Adding Integers
For use with Activity 11.2

Essential Question Is the sum of two integers *positive*, *negative*, or *zero*? How can you tell?

1 ACTIVITY: Adding Integers with the Same Sign

Work with a partner. Draw a picture to show how you use integer counters to find $-4 + (-3)$.

$-4 + (-3) =$ _____

2 ACTIVITY: Adding Integers with Different Signs

Work with a partner. Draw a picture to show how you use integer counters to find $-3 + 2$.

$-3 + 2 =$ _____

3 ACTIVITY: Adding Integers with Different Signs

Work with a partner. Show how to use a number line to find $5 + (-3)$.

$5 + (-3) =$ _____

11.2 Adding Integers (continued)

---

4 **ACTIVITY:** Adding Integers with Different Signs

Work with a partner. Write the addition expression shown. Then find the sum. How are the integers in the expression related to 0 on a number line?

Inductive Reasoning

Work with a partner. Use integer counters or a number line to complete the table.

	Exercise	Type of Sum	Sum	Sum: Positive, Negative, or Zero
1	**5.** $-4 + (-3)$			
2	**6.** $-3 + 2$			
3	**7.** $5 + (-3)$			
4	**8.** $7 + (-7)$			
	9. $2 + 4$			
	10. $-6 + (-2)$			
	11. $-5 + 9$			
	12. $15 + (-9)$			
	13. $-10 + 10$			
	14. $-6 + (-6)$			
	15. $13 + (-13)$			

11.2 **Adding Integers** (continued)

What Is Your Answer?

16. **IN YOUR OWN WORDS** Is the sum of two integers *positive*, *negative*, or *zero*? How can you tell?

17. **STRUCTURE** Write a general rule for adding

    a. two integers with the same sign.

    b. two integers with different signs.

    c. two integers that vary in sign.

Name _____ Date _____

11.2 Practice
For use after Lesson 11.2

Add.

1. $-9 + 2$

2. $5 + (-5)$

3. $-12 + (-6)$

4. $-10 + 19 + 5$

5. $-11 + (-20) + 9$

6. $-7 + 7 + (-8)$

Use mental math to solve the equation.

7. $x + (-5) = 4$

8. $y + 6 = -2$

9. $-10 = -7 + z$

10. The table shows the change in your hair length over a year.

Month	January	February	August	September	December
Change in hair length (inches)	2	−1	3	−4	3

   a. What is the total change in your hair length at the end of the year?

   b. Is your hair longer in January or December? Explain your reasoning.

   c. When is your hair the longest? Explain your reasoning.

Name_____ Date_____

 11.3 **Subtracting Integers**
For use with Activity 11.3

Essential Question How are adding integers and subtracting integers related?

1 **ACTIVITY:** Subtracting Integers

Work with a partner. Draw a picture to show how you use integer counters to find 4 − 2.

$4 - 2 = $ _____

2 **ACTIVITY:** Adding Integers

Work with a partner. Draw a picture to show how you use integer counters to find 4 + (−2).

$4 + (-2) = $ _____

3 **ACTIVITY:** Subtracting Integers

Work with a partner. Show how to use a number line to find −3 − 1.

$-3 - 1 = $ _____

11.3 **Subtracting Integers** (continued)

4 **ACTIVITY:** Adding Integers

Work with a partner. Write the addition expression shown. Then find the sum.

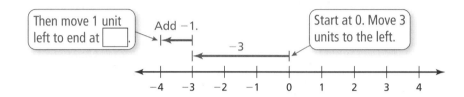

Inductive Reasoning

Work with a partner. Use integer counters or a number line to complete the table.

	Exercise	Operation: Add or Subtract	Answer
1	**5.** $4 - 2$		
2	**6.** $4 + (-2)$		
3	**7.** $-3 - 1$		
4	**8.** $-3 + (-1)$		
	9. $3 - 8$		
	10. $3 + (-8)$		
	11. $9 - 13$		
	12. $9 + (-13)$		
	13. $-6 - (-3)$		
	14. $-6 + 3$		
	15. $-5 - (-12)$		
	16. $-5 + 12$		

11.3 **Subtracting Integers** (continued)

What Is Your Answer?

17. IN YOUR OWN WORDS How are adding integers and subtracting integers related?

18. STRUCTURE Write a general rule for subtracting integers.

19. Use a number line to find the value of the expression $-4 + 4 - 9$. What property can you use to make your calculation easier? Explain.

11.3 Practice

For use after Lesson 11.3

Subtract.

1. $3 - 8$

2. $6 - (-7)$

3. $-10 - 9$

4. $-5 - (-4)$

Evaluate the expression.

5. $11 - (-2) + 14$

6. $-16 - (-12) + (-8)$

7. $6 - 17 - 4$

Use mental math to solve the equation.

8. $6 - x = 10$

9. $y - (-10) = 2$

10. $z - 17 = -14$

11. You begin a hike in Death Valley, California, at an elevation of -86 meters. You hike to a point of elevation at 45 meters. What is your change in elevation?

12. You sell T-shirts for a fundraiser. It costs \$112 to have the T-shirts made. You make \$98 in sales. What is your profit?

11.4 Multiplying Integers
For use with Activity 11.4

Essential Question Is the product of two integers *positive*, *negative*, or *zero*? How can you tell?

1 ACTIVITY: Multiplying Integers with the Same Sign

Work with a partner. Use repeated addition to find 3 • 2.

Recall that multiplication is repeated addition. 3 • 2 means to add 3 groups of 2.

$3 • 2 =$ _____

2 ACTIVITY: Multiplying Integers with Different Signs

Work with a partner. Use repeated addition to find 3 • (−2).

$3 • (−2) =$ _____

3 ACTIVITY: Multiplying Integers with Different Signs

Work with a partner. Use a table to find −3 • 2.

Describe the pattern of the products in the table. Then complete the table.

2	•	2	=	
1	•	2	=	
0	•	2	=	
−1	•	2	=	
−2	•	2	=	
−3	•	2	=	

$−3 • 2 =$ _____

11.4 **Multiplying Integers** (continued)

4 **ACTIVITY:** Multiplying Integers with the Same Sign

Work with a partner. Use a table to find $-3 \cdot (-2)$.

Describe the pattern of the products in the table. Then complete the table.

-3	$\cdot$	3	$=$	
-3	$\cdot$	2	$=$	
-3	$\cdot$	1	$=$	
-3	$\cdot$	0	$=$	
-3	$\cdot$	-1	$=$	
-3	$\cdot$	-2	$=$	

$-3 \cdot (-2) =$ _____

Inductive Reasoning

Work with a partner. Complete the table.

	Exercise	Type of Product	Product	Product: Positive or Negative
1	**5.** $3 \cdot 2$			
2	**6.** $3 \cdot (-2)$			
3	**7.** $-3 \cdot 2$			
4	**8.** $-3 \cdot (-2)$			
	9. $6 \cdot 3$			
	10. $2 \cdot (-5)$			
	11. $-6 \cdot 5$			
	12. $-5 \cdot (-3)$			

11.4 **Multiplying Integers** (continued)

What Is Your Answer?

13. Write two integers whose product is 0.

14. **IN YOUR OWN WORDS** Is the product of two integers *positive*, *negative*, or *zero*? How can you tell?

15. **STRUCTURE** Write a general rule for multiplying

   a. two integers with the same sign.

   b. two integers with different signs.

11.4 Practice
For use after Lesson 11.4

Multiply.

1. $8 \cdot 9$

2. $7(-7)$

3. $-10 \cdot 4$

4. $-5(-6)$

5. $12 \cdot (-1) \cdot (-2)$

6. $-10(-3)(-7)$

7. $-20 \cdot 0 \cdot (-4)$

8. $-4 \cdot 8 \cdot 3$

Evaluate the expression.

9. $(-8)^2$

10. -11^2

11. $9 \cdot (-5)^2$

12. $(-2)^3 \cdot (-6)$

13. You lose 5 points for every wrong answer in a trivia game. What integer represents the change in your points after answering 8 questions wrong?

11.5 Dividing Integers
For use with Activity 11.5

Essential Question Is the quotient of two integers *positive*, *negative*, or *zero*? How can you tell?

1 ACTIVITY: Dividing Integers with Different Signs

Work with a partner. Draw a picture to show how you use integer counters to find $-15 \div 3$.

$-15 \div 3 = $ _____

2 ACTIVITY: Rewriting a Product as a Quotient

Work with a partner. Rewrite the product $3 \cdot 4 = 12$ as a quotient in two different ways.

First Way

12 is equal to 3 groups of _____.

$12 \div 3 = $ _____

Second Way

12 is equal to 4 groups of _____.

$12 \div 4 = $ _____

3 ACTIVITY: Dividing Integers with Different Signs

Work with a partner. Rewrite the product $-3 \cdot (-4) = 12$ as a quotient in two different ways. What can you conclude?

First Way

Second Way

11.5 **Dividing Integers** (continued)

4 **ACTIVITY:** Dividing Negative Integers

Work with a partner. Rewrite the product $3 \cdot (-4) = -12$ as a quotient in two different ways. What can you conclude?

First Way   *Second Way*

Inductive Reasoning

Work with a partner. Complete the table.

	Exercise	Type of Quotient	Quotient	Quotient: Positive, Negative, or Zero
1	**5.** $-15 \div 3$			
2	**6.** $12 \div 4$			
3	**7.** $12 \div (-3)$			
4	**8.** $-12 \div (-4)$			
	9. $-6 \div 2$			
	10. $-21 \div (-7)$			
	11. $10 \div (-2)$			
	12. $12 \div (-6)$			
	13. $0 \div (-15)$			
	14. $0 \div 4$			

11.5 **Dividing Integers** (continued)

What Is Your Answer?

15. **IN YOUR OWN WORDS** Is the quotient of two integers *positive*, *negative*, or *zero*? How can you tell?

16. **STRUCTURE** Write a general rule for dividing

   a. two integers with the same sign.

   b. two integers with different signs.

11.5 Practice
For use after Lesson 11.5

Divide, if possible.

1. $3 \div (-1)$

2. $8 \div 2$

3. $-10 \div 5$

4. $-21 \div (-7)$

5. $\dfrac{48}{-6}$

6. $\dfrac{-13}{-13}$

7. $\dfrac{0}{3}$

8. $\dfrac{-55}{11}$

Evaluate the expression.

9. $-63 \div (-7) + 6$

10. $-5 - 12 \div 3$

11. $-8 \bullet 7 + 33 \div (-11)$

12. The table shows the number of yards a football player runs in each quarter of a game. Find the mean number of yards the player runs per quarter.

Quarter	1	2	3	4
Yards	−2	14	−18	−6

Chapter 12 **Fair Game Review**

Write the decimal as a fraction.

1. 0.26

2. 0.79

3. 0.571

4. 0.846

Write the fraction as a decimal.

5. $\dfrac{3}{8}$

6. $\dfrac{4}{10}$

7. $\dfrac{11}{16}$

8. $\dfrac{17}{20}$

9. A quarterback completed 0.6 of his passes during a game. Write the decimal as a fraction.

Chapter 12 Fair Game Review (continued)

Evaluate the expression.

10. $\dfrac{1}{8} + \dfrac{1}{9}$

11. $\dfrac{2}{3} + \dfrac{9}{10}$

12. $\dfrac{7}{12} - \dfrac{1}{4}$

13. $\dfrac{6}{7} - \dfrac{4}{5}$

14. $\dfrac{5}{9} \bullet \dfrac{1}{3}$

15. $\dfrac{8}{15} \bullet \dfrac{3}{4}$

16. $\dfrac{7}{8} \div \dfrac{11}{16}$

17. $\dfrac{3}{10} \div \dfrac{2}{5}$

18. You have 8 cups of flour. A recipe calls for $\dfrac{2}{3}$ cup of flour. Another recipe calls for $\dfrac{1}{4}$ cup of flour. How much flour do you have left after making the recipes?

Name_____ Date_____

12.1 Rational Numbers
For use with Activity 12.1

Essential Question How can you use a number line to order rational numbers?

A **rational number** is a number that can be written as a ratio of two integers.

$$2 = \frac{2}{1} \qquad\qquad -3 = \frac{-3}{1} \qquad\qquad -\frac{1}{2} = \frac{-1}{2} \qquad\qquad 0.25 = \frac{1}{4}$$

1 ACTIVITY: Ordering Rational Numbers

Work in groups of five. Order the numbers from least to greatest.

- Use masking tape and a marker to make a number line on the floor similar to the one shown.

- Write the numbers on pieces of paper. Then each person should choose one piece of paper.

- Stand on the location of your number on the number line.

- Use your positions to order the numbers from least to greatest.

  The numbers from least to greatest are

  _____, _____, _____, _____, and _____.

a. $-0.5, 1.25, -\frac{1}{3}, 0.5, -\frac{5}{3}$

b. $-\frac{7}{4}, 1.1, \frac{1}{2}, -\frac{1}{10}, -1.3$

c. $-1.4, -\frac{3}{5}, \frac{9}{2}, \frac{1}{4}, 0.9$

d. $\frac{5}{4}, 0.75, -\frac{5}{4}, -0.8, -1.1$

12.1 **Rational Numbers** (continued)

2 **ACTIVITY:** The Game of Math Card War

Preparation:

- Cut index cards to make 40 playing cards.*

- Write each number in the table on a card.

$\dfrac{3}{2}$	$\dfrac{3}{10}$	$-\dfrac{3}{4}$	-0.6	1.25	-0.15	$\dfrac{5}{4}$	$\dfrac{3}{5}$	-1.6	-0.3
$\dfrac{3}{20}$	$\dfrac{8}{5}$	-1.2	$\dfrac{19}{10}$	0.75	-1.5	$-\dfrac{6}{5}$	$-\dfrac{3}{5}$	1.2	0.3
1.5	1.9	-0.75	-0.4	$\dfrac{3}{4}$	$-\dfrac{5}{4}$	-1.9	$\dfrac{2}{5}$	$-\dfrac{3}{20}$	$-\dfrac{19}{10}$
$\dfrac{6}{5}$	$-\dfrac{3}{10}$	1.6	$-\dfrac{2}{5}$	0.6	0.15	$\dfrac{3}{2}$	-1.25	0.4	$-\dfrac{8}{5}$

To Play:

- Play with a partner.

- Deal 20 cards to each player facedown.

- Each player turns one card faceup. The player with the greater number wins. The winner collects both cards and places them at the bottom of his or her cards.

- Suppose there is a tie. Each player lays three cards facedown, then a new card faceup. The player with the greater of these new cards wins. The winner collects all ten cards and places them at the bottom of his or her cards.

- Continue playing until one player has all the cards. This player wins the game.

*Cut-outs are available in the back of the Record and Practice Journal.

12.1 Rational Numbers (continued)

What Is Your Answer?

3. **IN YOUR OWN WORDS** How can you use a number line to order rational numbers? Give an example.

The numbers are in order from least to greatest. Fill in the blank spaces with rational numbers.

4. $-\dfrac{1}{2}$, ☐ , $\dfrac{1}{3}$, ☐ , $\dfrac{7}{5}$, ☐

5. $-\dfrac{5}{2}$, ☐ , -1.9, ☐ , $-\dfrac{2}{3}$, ☐

6. $-\dfrac{1}{3}$, ☐ , -0.1, ☐ , $\dfrac{4}{5}$, ☐

7. -3.4, ☐ , -1.5, ☐ , 2.2, ☐

Name _____ Date _____

<table>
<tr><td>**12.1**</td><td>**Practice**
For use after Lesson 12.1</td></tr>
</table>

Write the rational number as a decimal.

1. $-\dfrac{9}{10}$

2. $-4\dfrac{2}{3}$

3. $1\dfrac{7}{16}$

Write the decimal as a fraction or mixed number in simplest form.

4. -0.84

5. 5.22

6. -1.716

Order the numbers from least to greatest.

7. $\dfrac{1}{5}, 0.1, -\dfrac{1}{2}, -0.25, 0.3$

8. $-1.6, \dfrac{5}{2}, -\dfrac{7}{8}, 0.9, -\dfrac{6}{5}$

9. $-\dfrac{2}{3}, \dfrac{5}{9}, 0.5, -1.3, -\dfrac{10}{3}$

10. The table shows the position of each runner relative to when the first place finisher crossed the finish line. Who finished in second place? Who finished in fifth place?

Runner	A	B	C	D	E	F
Meters	-1.264	$-\dfrac{5}{4}$	-1.015	-0.480	$-\dfrac{14}{25}$	$-\dfrac{13}{8}$

12.2 **Adding Rational Numbers**
For use with Activity 12.2

Essential Question How can you use what you know about adding integers to add rational numbers?

---

1 **ACTIVITY:** Adding Rational Numbers

Work with a partner. Use a number line to find the sum.

a. $2.7 + (-3.4)$

$2.7 + (-3.4) =$ _____

b. $1.3 + (-1.5)$ **c.** $-2.1 + 0.8$

d. $-1\dfrac{1}{4} + \dfrac{3}{4}$ **e.** $\dfrac{3}{10} + \left(-\dfrac{3}{10}\right)$

---

2 **ACTIVITY:** Adding Rational Numbers

Work with a partner. Use a number line to find the sum.

a. $-1\dfrac{2}{5} + \left(-\dfrac{4}{5}\right)$

$-1\dfrac{2}{5} + \left(-\dfrac{4}{5}\right) =$ _____

12.2 **Adding Rational Numbers** (continued)

b. $-\dfrac{7}{10} + \left(-1\dfrac{7}{10}\right)$ **c.** $-1\dfrac{2}{3} + \left(-1\dfrac{1}{3}\right)$

d. $-0.4 + (-1.9)$ **e.** $-2.3 + (-0.6)$

3 **ACTIVITY:** Writing Expressions

Work with a partner. Write the addition expression shown. Then find the sum.

a.

b.

12.2   **Adding Rational Numbers** (continued)

c.

Then move 0.7 unit left to end at ☐.   Add −0.7.   Start at 0. Move 1.1 units to the left.

What Is Your Answer?

4. **IN YOUR OWN WORDS** How can you use what you know about adding integers to add rational numbers?

PUZZLE Find a path through the table so that the numbers add up to the sum. You can move horizontally or vertically.

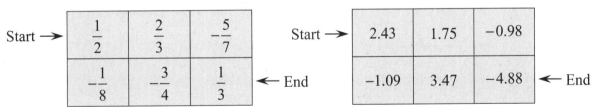

5. Sum: $\dfrac{3}{4}$

Start →

$\dfrac{1}{2}$	$\dfrac{2}{3}$	$-\dfrac{5}{7}$
$-\dfrac{1}{8}$	$-\dfrac{3}{4}$	$\dfrac{1}{3}$

← End

6. Sum: −0.07

Start →

2.43	1.75	−0.98
−1.09	3.47	−4.88

← End

12.2 Practice
For use after Lesson 12.2

Add. Write fractions in simplest form.

1. $-\dfrac{4}{5} + \dfrac{3}{20}$

2. $-8 + \left(-\dfrac{6}{7}\right)$

3. $1\dfrac{2}{15} + \left(-3\dfrac{1}{2}\right)$

4. $-\dfrac{1}{6} + \left(-\dfrac{5}{12}\right)$

5. $\dfrac{9}{10} + (-3)$

6. $-5\dfrac{3}{4} + \left(-4\dfrac{5}{6}\right)$

7. $0.46 + (-0.642)$

8. $0.13 + (-5.7)$

9. $-2.57 + (-3.48)$

10. Before a race, you start $4\dfrac{5}{8}$ feet behind your friend. At the halfway point, you are $3\dfrac{2}{3}$ feet ahead of your friend. What is the change in distance between you and your friend from the beginning of the race?

12.3 Subtracting Rational Numbers
For use with Activity 12.3

Essential Question How can you use what you know about subtracting integers to subtract rational numbers?

1 ACTIVITY: Subtracting Rational Numbers

Work with a partner. Use a number line to find the difference.

a. $-1\dfrac{1}{2} - \dfrac{1}{2}$

$-1\dfrac{1}{1} - \dfrac{1}{2} =$ _____

b. $\dfrac{6}{10} - 1\dfrac{3}{10}$

c. $-1\dfrac{1}{4} - 1\dfrac{3}{4}$

d. $-1.9 - 0.8$

e. $0.2 - 0.7$

2 ACTIVITY: Finding Distances on a Number Line

Work with a partner.

a. Plot -3 and 2 on the number line. Then find $-3 - 2$ and $2 - (-3)$. What do you notice about your results?

12.3 **Subtracting Rational Numbers** (continued)

b. Plot $\frac{3}{4}$ and 1 on the number line. Then find $\frac{3}{4} - 1$ and $1 - \frac{3}{4}$. What do you notice about your results?

c. Choose any two points a and b on a number line. Find the values of $a - b$ and $b - a$. What do the absolute values of these differences represent? Is this true for any pair of rational numbers? Explain.

---

3 **ACTIVITY:** Financial Literacy

Work with a partner. The table shows the balance in a checkbook.

- Deposits and interest are amounts added to the account.

- Amounts shown in parentheses are taken from the account.

Date	Check #	Transaction	Amount	Balance
--	--	Previous Balance	--	100.00
1/02/2013	124	Groceries	(34.57)	
1/07/2013		Check deposit	875.50	
1/11/2013		ATM withdrawal	(40.00)	
1/14/2013	125	Electric company	(78.43)	
1/17/2013		Music store	(10.55)	
1/18/2013	126	Shoes	(47.21)	
1/22/2013		Check deposit	125.00	
1/24/2013		Interest	2.12	
1/25/2013	127	Cell phone	(59.99)	
1/26/2013	128	Clothes	(65.54)	
1/30/2013	129	Cable company	(75.00)	

12.3 **Subtracting Rational Numbers** (continued)

You can find the balance in the second row two different ways.

$100.00 - 34.57 = 65.43$ Subtract 34.57 from 100.00.

$100.00 + (-34.57) = 65.43$ Add –34.57 to 100.00.

   a. Complete the balance column of the table on the previous page.

   b. How did you find the balance in the twelfth row?

   c. Use a different way to find the balance in part (b).

What Is Your Answer?

   4. IN YOUR OWN WORDS How can you use what you know about subtracting integers to subtract rational numbers?

   5. Give two real-life examples of subtracting rational numbers that are not integers.

Name _____ Date _____

Subtract. Write fractions in simplest form.

1. $\dfrac{4}{9} - \left(-\dfrac{2}{9}\right)$

2. $-2\dfrac{3}{7} - 1\dfrac{2}{3}$

3. $-2.35 - (-1.27)$

Find the distance between the two numbers on a number line.

4. $-3\dfrac{1}{4},\ -6\dfrac{1}{2}$

5. $-1.5,\ 2.8$

6. $-4,\ -7\dfrac{1}{3}$

Evaluate.

7. $2\dfrac{1}{2} + \left(-\dfrac{7}{6}\right) - 1\dfrac{3}{4}$

8. $2.37 - (-1.55) - 2.48$

9. Your friend drinks $\dfrac{2}{3}$ of a bottle of water. You drink $\dfrac{5}{7}$ of a bottle of water. Find the difference of the amounts of water left in each bottle.

12.4 Multiplying and Dividing Rational Numbers
For use with Activity 12.4

Essential Question Why is the product of two negative rational numbers positive?

---

1 ACTIVITY: Showing $(-1)(-1) = 1$

Work with a partner. How can you show that $(-1)(-1) = 1$?

To begin, assume that $(-1)(-1) = 1$ is a true statement. From the Additive Inverse Property, you know that $1 + (-1) = 0$. So, substitute $(-1)(-1)$ for 1 to get $(-1)(-1) + (-1) = 0$. If you can show that $(-1)(-1) + (-1) = 0$ is true, then you have shown that $(-1)(-1) = 1$.

Justify each step.

$$(-1)(-1) + (-1) = (-1)(-1) + 1(-1) \quad \rule{6cm}{0.4pt}$$

$$= (-1)\left[(-1) + 1\right] \quad \rule{6cm}{0.4pt}$$

$$= (-1)0 \quad \rule{6cm}{0.4pt}$$

$$= 0 \quad \rule{6cm}{0.4pt}$$

$$(-1)(-1) = \rule{2cm}{0.4pt}$$

---

2 ACTIVITY: Multiplying by −1

Work with a partner.

a. Graph each number below on three different number lines. Then multiply each number by –1 and graph the product on the appropriate number line.

| 2 | 8 | −1 |

12.4 **Multiplying and Dividing Rational Numbers** (continued)

b. How does multiplying by −1 change the location of the points in part (a)? What is the relationship between the number and the product?

c. Graph each number below on three different number lines. Where do you think the points will be after multiplying by −1? Plot the points. Explain your reasoning.

$$\frac{1}{2} \qquad 2.5 \qquad -\frac{5}{2}$$

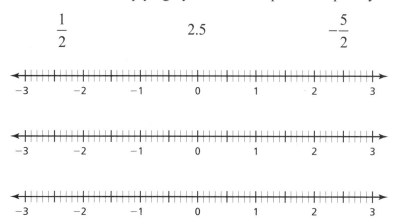

d. What is the relationship between a rational number −a and the product −1(a)? Explain your reasoning.

3 **ACTIVITY: Understanding the product of Rational Numbers**

Work with a partner. Let a and b be positive rational numbers.

a. Because a and b are positive, what do you know about −a and −b?

b. Justify each step.

$$(-a)(-b) = (-1)(a)(-1)(b) \quad \rule{3cm}{0.4pt}$$
$$= (-1)(-1)(a)(b) \quad \rule{3cm}{0.4pt}$$
$$= (1)(a)(b) \quad \rule{3cm}{0.4pt}$$
$$= ab \quad \rule{3cm}{0.4pt}$$

c. Because a and b are positive, what do you know about the product ab?

12.4 Multiplying and Dividing Rational Numbers (continued)

d. What does this tell you about products of rational numbers? Explain.

4 **ACTIVITY:** Writing a Story

Work with a partner. Write a story that uses addition, subtraction, multiplication, or division of rational numbers.

- At least one of the numbers in the story has to be negative and *not* an integer.

- Draw pictures to help illustrate what is happening in the story.

- Include the solution of the problem in the story.

If you are having trouble thinking of a story, here are some common uses of negative numbers:

- A profit of −$15 is a loss of $15.

- An elevation of −100 feet is a depth of 100 feet below sea level.

- A gain of −5 yards in football is a loss of 5 yards.

- A score of −4 in golf is 4 strokes under par.

What Is Your Answer?

5. IN YOUR OWN WORDS Why is the product of two negative rational numbers positive?

6. PRECISION Show that $(-2)(-3) = 6$.

7. How can you show that the product of a negative rational number and a positive rational number is negative?

12.4 Practice
For use after Lesson 12.4

Multiply or divide. Write fractions in simplest form.

1. $-\dfrac{8}{9}\left(-\dfrac{18}{25}\right)$

2. $-4\left(\dfrac{9}{16}\right)$

3. $-3\dfrac{3}{7} \times 2\dfrac{1}{2}$

4. $-\dfrac{2}{3} \div \dfrac{5}{9}$

5. $\dfrac{7}{13} \div (-2)$

6. $-5\dfrac{5}{8} \div \left(-4\dfrac{7}{12}\right)$

7. $-1.39 \times (-6.8)$

8. $-10 \div 0.22$

9. $-12.166 \div (-1.54)$

10. In a game of tug of war, your team changes $-1\dfrac{3}{10}$ feet in position every 10 seconds. What is your change in position after 30 seconds?

Evaluate the expression when $x = \dfrac{1}{2}$ **and** $y = -3$.

1. $-4xy$

2. $6x - 3y$

3. $-5y + 8x + 1$

4. $-x^2 - y + 2$

5. $2x + 4\left(x + \dfrac{1}{2}\right) + 7$

6. $2(4x - 1)^2 - 3$

7. Find the area of the garden when $x = 2$ feet.

3x ft

2x ft

Chapter 13 **Fair Game Review** (continued)

Write the phrase as an algebraic expression.

8. the sum of eight and a number y

9. six less than a number p

10. the product of seven and a number m

11. eight less than the product of eleven and a number c

12. a number r decreased by the quotient of a number r and two

13. the product of nine and the sum of a number z and four

13.1 Simplifying Algebraic Expressions
For use with Activity 13.1

Essential Question How can you simplify an algebraic expression?

---

1 ACTIVITY: Simplifying Algebraic Expressions

Work with a partner.

a. Evaluate each algebraic expression when $x = 0$ and when $x = 1$. Use the results to match each expression in the left table with its equivalent expression in the right table.

	Expression	Value When $x = 0$	Value When $x = 1$
A.	$3x + 2 - x + 4$		
B.	$5(x - 3) + 2$		
C.	$x + 3 - (2x + 1)$		
D.	$-4x + 2 - x + 3x$		
E.	$-(1 - x) + 3$		
F.	$2x + x - 3x + 4$		
G.	$4 - 3 + 2(x - 1)$		
H.	$2(1 - x + 4)$		
I.	$5 - (4 - x + 2x)$		
J.	$5x - (2x + 4 - x)$		

	Expression	Value When $x = 0$	Value When $x = 1$
a.	4		
b.	$-x + 1$		
c.	$4x - 4$		
d.	$2x + 6$		
e.	$5x - 13$		
f.	$-2x + 10$		
g.	$x + 2$		
h.	$2x - 1$		
i.	$-2x + 2$		
j.	$-x + 2$		

b. Compare each expression in the left table with its equivalent expression in the right table. In general, how do you think you obtain the equivalent expression in the right table?

13.1 **Algebraic Expressions** (continued)

2 **ACTIVITY:** Writing a Math Lesson

Work with a partner. Use your results from Activity 1 to write a lesson on simplifying an algebraic expression.

Simplifying an Algebraic Expression

Key Idea Use the following steps to simplify an algebraic expression.

> Describe steps you can use to simplify an expression.

   1.

   2.

   3.

Examples ← Write 3 examples. Use expressions from Activity 1.

   a.

   b.

   c.

Exercises ← Write 3 exercises. Use expressions different from the ones in Activity 1.

Simplify the expression.

   1.

   2.

   3.

13.1 Algebraic Expressions (continued)

What Is Your Answer?

3. **IN YOUR OWN WORDS** How can you simplify an algebraic expression?
   Give an example that demonstrates your procedure.

4. **REASONING** Why would you want to simplify an algebraic expression?
   Discuss several reasons.

13.1 Practice
For use after Lesson 13.1

Identify the terms and like terms in the expression.

1. $3x + 4 - 7x - 6$

2. $-9 + 2.5y - 0.7y + 1 + 6.4y^2$

Simplify the expression.

3. $5a - 2a + 9$

4. $m - \dfrac{1}{6} - 4m + \dfrac{5}{6}$

5. $2.3w - 7 + 8.1 - 3w$

6. $7(d - 1) + 2$

7. $13g + 2(4k - g)$

8. $20(p + 2) + 16(-3 - p)$

9. Write an expression in simplest form that represents the cost for shampooing and cutting w women's hair and m men's hair.

	Women	Men
Cut	$15	$7
Shampoo	$5	$2

13.2 Adding and Subtracting Linear Expressions

For use with Activity 13.2

Essential Question How can you use algebra tiles to add or subtract algebraic expressions?

Key: [+] = variable   [−] = −variable   [+][−] = zero pair

[+] = 1   [−] = −1   [+][−] = zero pair

1 ACTIVITY: Writing Algebraic Expressions

Work with a partner. Write an algebraic expression shown by the algebra tiles.

a. [+] [+][+][+]

b. [+] [−][−]
   [+]

c. [+] [+][+][+][+]
   [+] [−][−]

d. [+] [+][+][+]
   [+] [−][−][−][−]
   [+] [−][−]

2 ACTIVITY: Adding Algebraic Expressions

Work with a partner. Write the sum of two algebraic expressions modeled by the algebra tiles. Then use the algebra tiles to simplify the expression.

a. ([+] [+][+]) + ([+] [+][+][+][+])

b. ([+] [−][−][−][−][−]) + ([+] [−][−])

13.2 Adding and Subtracting Linear Expressions (continued)

c. $\left(\boxed{+} \; \boxed{+}\,\boxed{+}\,\boxed{+}\,\boxed{+} \right) + \left(\boxed{\begin{array}{c}+\\+\end{array}} \; \boxed{-}\,\boxed{-}\,\boxed{-} \right)$

d. $\left(\boxed{\begin{array}{c}+\\+\end{array}} \; \boxed{-}\,\boxed{-}\,\boxed{-}\,\boxed{-}\,\boxed{-}\;\boxed{-}\,\boxed{-}\,\boxed{-} \right) + \left(\boxed{\begin{array}{c}+\\+\\+\end{array}} \; \boxed{+}\,\boxed{+}\,\boxed{+}\;\boxed{+}\,\boxed{+} \right)$

---

3 **ACTIVITY:** Subtracting Algebraic Expressions

Work with a partner. Write the difference of two algebraic expressions modeled by the algebra tiles. Then use the algebra tiles to simplify the expression.

a. $\left(\boxed{+} \; \boxed{+}\,\boxed{+}\,\boxed{+} \right) - \left(\boxed{+} \; \boxed{+} \right)$

b. $\left(\boxed{+} \; \boxed{-}\,\boxed{-}\,\boxed{-}\,\boxed{-} \right) - \left(\boxed{+} \; \boxed{-}\,\boxed{-}\,\boxed{-} \right)$

c. $\left(\boxed{\begin{array}{c}+\\+\end{array}} \; \boxed{+}\,\boxed{+}\,\boxed{+}\,\boxed{+}\,\boxed{+} \right) - \left(\boxed{+} \; \boxed{-} \right)$

13.2 **Adding and Subtracting Linear Expressions** (continued)

d. $\left(\begin{array}{c} \boxed{+} \\ \boxed{+} \\ \boxed{+} \end{array} \begin{array}{ccccc} \blacksquare & \blacksquare & \blacksquare & \blacksquare & \blacksquare \\ \blacksquare & \blacksquare & & & \end{array} \right) - \left(\begin{array}{c} \boxed{+} \\ \boxed{+} \end{array} \begin{array}{ccc} \boxed{+} & \boxed{+} & \boxed{+} \end{array} \right)$

---

4 **ACTIVITY:** Adding and Subtracting Algebraic Expressions

Work with a partner. Use algebra tiles to model the sum or difference. Then use the algebra tiles to simplify the expression.

a. $(2x + 1) + (x - 1)$

b. $(2x - 6) + (3x + 2)$

c. $(2x + 4) - (x + 2)$

d. $(4x + 3) - (2x - 1)$

What Is Your Answer?

5. **IN YOUR OWN WORDS** How can you use algebra tiles to add or subtract algebraic expressions?

6. Write the difference of two algebraic expressions modeled by the algebra tiles. Then use the algebra tiles to simplify the expression.

$\left(\begin{array}{cccc} \blacksquare & \boxed{+} & \boxed{+} & \boxed{+} \end{array} \right) - \left(\begin{array}{ccc} \blacksquare & \blacksquare & \blacksquare \\ \blacksquare & & \blacksquare \end{array} \right)$

13.2 Practice
For use after Lesson 13.2

Find the sum or difference.

1. $(x - 2) + (x + 6)$

2. $(2n - 4) - (4n - 3)$

3. $2(-3y - 1) + (2y + 7)$

4. $(1 - 3k) - 4(2 + 2.5k)$

5. $(6g - 9) + \dfrac{1}{3}(15 - 9g)$

6. $\dfrac{1}{2}(2r + 4) - \dfrac{1}{4}(16 - 8r)$

7. You earn $(4x + 12)$ points after completing x levels of a video game and then lose $(2x - 5)$ points. Write an expression that represents the total number of points you have now.

Extension 13.2 **Practice**
For use after Extension 13.2

Factor the expression using the GCF.

1. $7 + 28$

2. $25 + 50$

3. $7b - 7$

4. $8a - 16$

5. $8x + 12$

6. $12y + 24t$

7. $10w + 50z$

8. $10v + 12u$

9. $9a + 15b$

Factor out the coefficient of the variable.

10. $\dfrac{1}{2}a - \dfrac{1}{2}$

11. $\dfrac{1}{4}d - \dfrac{3}{4}$

12. $\dfrac{5}{6}s + \dfrac{2}{3}$

**Extension
13.2** **Practice** (continued)

Factor out the coefficient of the variable.

13. $\dfrac{3}{10}y - \dfrac{2}{5}$

14. $1.1x + 9.9$

15. $3.4c + 10.2$

16. Factor -2 out of $-6x + 10$.

17. Factor $-\dfrac{1}{3}$ out of $\dfrac{1}{3}y - \dfrac{3}{2}$.

18. A square window has a perimeter of $(8x + 12)$ feet. Write an expression that represents the side length of the window (in feet).

13.3 Solving Equations Using Addition or Subtraction
For use with Activity 13.3

Essential Question How can you use algebra tiles to solve addition or subtraction equations?

1 ACTIVITY: Solving Equations

Work with a partner. Use algebra tiles to model and solve the equation.

a. $x - 3 = -4$

Model the equation $x - 3 = -4$. Draw a sketch of your tiles.

To get the variable tile by itself, remove the _____ tiles on the left side by adding _____ tiles to each side.

How many *zero pairs* can you remove from each side? Circle them.

The remaining tiles show the value of x.

$x =$ _____

b. $z - 6 = 2$

c. $p - 7 = -3$

d. $-15 = t - 5$

13.3 Solving Equations Using Addition or Subtraction (continued)

2 ACTIVITY: Solving Equations

Work with a partner. Use algebra tiles to model and solve the equation.

a. $-5 = n + 2$

b. $y + 10 = -5$

c. $7 + b = -1$

d. $8 = 12 + z$

3 ACTIVITY: Writing and Solving Equations

Work with a partner. Write an equation shown by the algebra tiles. Then solve.

a.

b.

c.

d.

13.3 Solving Equations Using Addition or Subtraction (continued)

4 **ACTIVITY:** Using a Different Method to Find a Solution

Work with a partner. The *melting point* of a solid is the temperature at which the solid melts to become a liquid. The melting point of the element bromine is about 19°F. This is about 57°F more than the melting point of mercury.

a. Which of the following equations can you use to find the melting point of mercury? What is the melting point of mercury?

| $x + 57 = 19$ | $x - 57 = 19$ | $x + 19 = 57$ | $x + 19 = -57$ |

b. **CHOOSE TOOLS** How can you solve this problem without using an equation? Explain. How are these two methods related?

What Is Your Answer?

5. **IN YOUR OWN WORDS** How can you use algebra tiles to solve addition or subtraction equations? Give an example of each.

6. **STRUCTURE** Explain how you could use inverse operations to solve addition or subtraction equations without using algebra tiles.

7. What makes the cartoon funny?

"**Dear Sir: Yesterday you said *x* = 2. Today you are saying *x* = 3. Please make up your mind.**"

8. The word *variable* comes from the word *vary*. For example, the temperature in Maine varies a lot from winter to summer. Write two other English sentences that use the word *vary*.

Name _____ Date _____

Solve the equation. Check your solution.

1. $y + 12 = -26$

2. $15 + c = -12$

3. $-16 = d + 21$

4. $n + 12.8 = -0.3$

5. $1\frac{1}{8} = g - 4\frac{2}{5}$

6. $-5.47 + k = -14.19$

Write the word sentence as an equation. Then solve.

7. 42 less than x is -50.

8. 32 is the sum of a number z and 9.

9. A clothing company makes a profit of $2.3 million. This is $4.1 million more than last year. What was the profit last year?

10. A drop on a wooden roller coaster is $-98\frac{1}{2}$ feet. A drop on a steel roller coaster is $100\frac{1}{4}$ feet lower than the drop on the wooden roller coaster. What is the drop on the steel roller coaster?

13.4 Solving Equations Using Multiplication or Division
For use with Activity 13.4

Essential Question How can you use multiplication or division to solve equations?

1 ACTIVITY: Using Division to Solve Equations

Work with a partner. Use algebra tiles to model and solve the equation.

a. $3x = -12$

Model the equation $3x = -12$. Draw a sketch of your tiles.

Your goal is to get one variable tile by itself. Because there are _____ variable tiles, divide the _____ tiles into _____ equal groups. Circle the groups.

Keep one of the groups. This shows the value of x. Draw a sketch of the remaining tiles.

$x =$ _____ .

b. $2k = -8$

13.4 **Solving Equations Using Multiplication or Division** (continued)

    c. $-15 = 3t$

    d. $-20 = 5m$

    e. $4h = -16$

2 **ACTIVITY:** Writing and Solving Equations

Work with a partner. Write an equation shown by the algebra tiles. Then solve.

a.

+ +	+
+ +	+
+ + =	+
+ +	+

b.

+ +	− − −
+ + =	− − −
+	− − −
+	− − −

c.

− −	
− −	+
− − =	+
− −	
− −	

d.

	− − − − −
+	− − − − −
+ =	− − − − −
+	− − −
	− − −

13.4 Solving Equations Using Multiplication or Division (continued)

3 ACTIVITY: Using a Different Method to Find a Solution

**Work with a partner. Choose the equation you can use to solve each problem.
Solve the equation. Then explain how to solve the problem without using an
equation. How are the two methods related?**

a. For the final part of a race, a handcyclist travels 32 feet each second across
a distance of 400 feet. How many seconds does it take for the handcyclist
to travel the last 400 feet of the race?

$32x = 400$	$400x = 32$	$\dfrac{x}{32} = 400$	$\dfrac{x}{400} = 32$

b. The melting point of the element radon is about $-96°F$. The melting point of
nitrogen is about 3.6 times the melting point of radon. What is the melting
point of nitrogen?

$3.6x = -96$	$x + 96 = 3.6$	$\dfrac{x}{3.6} = -96$	$-96x = 3.6$

c. This year, a hardware store has a profit of $-\$6.0$ million. This profit is $\dfrac{3}{4}$ of
last year's profit. What is last year's profit?

$\dfrac{x}{-6} = \dfrac{3}{4}$	$-6x = \dfrac{3}{4}$	$\dfrac{3}{4} + x = -6$	$\dfrac{3}{4}x = -6$

What Is Your Answer?

4. **IN YOUR OWN WORDS** How can you use multiplication or division to
solve equations? Give an example of each.

Name _____ Date _____

13.4 Practice
For use after Lesson 13.4

Solve the equation. Check your solution.

1. $\dfrac{d}{5} = -6$

2. $8x = -6$

3. $-15 = \dfrac{z}{-2}$

4. $3.2n = -0.8$

5. $-\dfrac{3}{10}h = 15$

6. $-1.1k = -1.21$

Write the word sentence as an equation. Then solve.

7. A number divided by −8 is 7.

8. The product of −12 and a number is 60.

9. You earn $0.85 for every cup of hot chocolate you sell. How many cups do you need to sell to earn $55.25?

Name_____ Date _____

13.5 Solving Two-Step Equations
For use with Activity 13.5

Essential Question How can you use algebra tiles to solve a two-step equation?

1 ACTIVITY: Solving a Two-Step Equation

Work with a partner. Use algebra tiles to model and solve $2x - 3 = -5$.

Model the equation $2x - 3 = -5$.
Draw a sketch of your tiles.

Remove the _____ red tiles on the left side by
adding _____ yellow tiles to each side.

How many *zero pairs* can you remove from each side?
Circle them.

Because there are _____ green tiles, divide
the red tiles into _____ equal groups.
Circle the groups.

Keep one of the groups. This shows the value of x.
Draw a sketch of the remaining tiles.

$x =$ _____.

2 ACTIVITY: The Math behind the Tiles

**Work with a partner. Solve $2x - 3 = -5$ without using algebra tiles.
Complete each step. Then answer the questions.**

   a. Which step is first, adding 3 to each side or dividing each side by 2?

   b. How are the above steps related to the steps in Activity 1?

13.5 **Solving Two-Step Equations** (continued)

3 **ACTIVITY:** Solving Equations Using Algebra Tiles

Work with a partner.

- **Write an equation shown by the algebra tiles.**

- **Use algebra tiles to model and solve the equation.**

- **Check your answer by solving the equation without using algebra tiles.**

a.

b.

4 **ACTIVITY:** Working Backwards

Work with a partner.

a. Your friend pauses a video game to get a drink. You continue the game. You double the score by saving a princess. Then you lose 75 points because you do not collect the treasure. You finish the game with –25 points. How many points did you have when you started?

One way to solve the problem is to work backwards. To do this, start with the end result and retrace the events.

You started the game with _____ points.

13.5 **Solving Two-Step Equations** (continued)

   b. You triple your account balance by making a deposit. Then you withdraw
      $127.32 to buy groceries. Your account is now overdrawn by $10.56.
      By working backwards, find your account balance before you made
      the deposit.

What Is Your Answer?

   5. **IN YOUR OWN WORDS** How can you use algebra tiles to solve a two-step
      equation?

   6. When solving the equation $4x + 1 = -11$, what is the first step?

   7. **REPEATED REASONING** Solve the equation $2x - 75 = -25$. How do
      your steps compare with the strategy of working backwards in Activity 4?

13.5 Practice
For use after Lesson 13.5

Solve the equation. Check your solution.

1. $3a - 5 = -14$

2. $10 = -2c + 22$

3. $18 = -5b - 17$

4. $-12 = -8z + 12$

5. $1.3n - 0.03 = -9$

6. $-\dfrac{5}{11}h + \dfrac{7}{9} = \dfrac{2}{9}$

7. The length of a rectangle is 3 meters less than twice its width.

   a. Write an equation to find the length of the rectangle.

   b. The length of the rectangle is 11 meters. What is the width of the rectangle?

Name_____ Date _____

Fair Game Review

Simplify.

1. $\dfrac{3}{18}$

2. $\dfrac{4}{6}$

3. $\dfrac{12}{60}$

4. $\dfrac{14}{28}$

5. $\dfrac{16}{36}$

6. $\dfrac{40}{50}$

Are the fractions equivalent?

7. $\dfrac{3}{8} \overset{?}{=} \dfrac{6}{11}$

8. $\dfrac{4}{10} \overset{?}{=} \dfrac{16}{40}$

9. $\dfrac{22}{32} \overset{?}{=} \dfrac{11}{16}$

10. $\dfrac{63}{72} \overset{?}{=} \dfrac{7}{9}$

11. You see 58 birds while on a bird watching tour. Of those birds, you see 12 hawks. Write and simplify the fraction of hawks you see.

Solve the equation. Check your solution.

12. $\dfrac{d}{12} = -4$

13. $-7 = \dfrac{x}{-3}$

14. $\dfrac{1}{8}n = 5$

15. $6a = -54$

16. $10 = -2k$

17. $2.7 = -0.9y$

18. $-23.4 = -1.3w$

19. $\dfrac{1}{15}z = 6$

20. You and three friends spend $35 on tickets at the movies. Write and solve an equation to find the price p of one ticket.

14.1 Ratios and Rates

For use with Activity 14.1

Essential Question How do rates help you describe real-life problems?

1 ACTIVITY: Finding Reasonable Rates

Work with a partner.

a. Match each description with a verbal rate.

b. Match each verbal rate with a numerical rate.

c. Give a reasonable numerical rate for each description. Then give an unreasonable rate.

Description	*Verbal Rate*	*Numerical Rate*
Your running rate in a 100-meter dash	Dollars per year	$= \dfrac{\boxed{}\text{ in.}}{\text{yr}}$
The fertilization rate for an apple orchard	Inches per year	$= \dfrac{\boxed{}\text{ lb}}{\text{acre}}$
The average pay rate for a professional athlete	Meters per second	$= \dfrac{\$\,\boxed{}}{\text{yr}}$
The average rainfall rate in a rainforest	Pounds per acre	$= \dfrac{\boxed{}\text{ m}}{\text{sec}}$

2 ACTIVITY: Simplifying Expressions That Contain Fractions

Work with a partner. Describe a situation where the given expression may apply. Show how you can rewrite each expression as a division problem. Then simplify and interpret your result.

a. $\dfrac{\frac{1}{2}\,\text{c}}{4\text{ fl oz}}$

b. $\dfrac{2\text{ in.}}{\frac{3}{4}\text{ sec}}$

14.1 **Ratios and Rates** (continued)

c. $\dfrac{\dfrac{3}{8} \text{ c sugar}}{\dfrac{3}{5} \text{ c flour}}$

d. $\dfrac{\dfrac{5}{6} \text{ gal}}{\dfrac{2}{3} \text{ sec}}$

---

3 **ACTIVITY:** Using Ratio Tables to Find Equivalent Rates

Work with a partner. A communications satellite in orbit travels about 18 miles every 4 seconds.

a. Identify the rate in this problem.

b. Recall that you can use *ratio tables* to find and organize equivalent ratios and rates. Complete the ratio table below.

Time (seconds)	4	8	12	16	20
Distance (miles)					

c. How can you use a ratio table to find the speed of the satellite in miles per minute? miles per hour?

d. How far does the satellite travel in 1 second? Solve this problem (1) by using a ratio table and (2) by evaluating a quotient.

e. How far does the satellite travel in $\dfrac{1}{2}$ second? Explain your steps.

Name_____ Date_____

4 **ACTIVITY:** Unit Analysis

Work with a partner. Describe a situation where the product may apply.
Then find each product and list the units.

a. $10 \text{ gal} \times \dfrac{22 \text{ mi}}{\text{gal}}$

b. $\dfrac{7}{2} \text{ lb} \times \dfrac{\$3}{\frac{1}{2}\text{lb}}$

c. $\dfrac{1}{2} \text{ sec} \times \dfrac{30 \text{ ft}^2}{\text{sec}}$

What Is Your Answer?

5. **IN YOUR OWN WORDS** How do rates help you describe real-life
   problems? Give two examples.

6. To estimate the annual salary for a given hourly pay rate, multiply by 2 and
   insert "000" at the end.

   Sample: $10 per hour is about $20,000 per year.

   a. Explain why this works. Assume the person is working 40 hours a week.

   b. Estimate the annual salary for an hourly pay rate of $8 per hour.

   c. You earn $1 million per month. What is your
   annual salary?

   d. Why is the cartoon funny?

"We had someone apply for
the job. He says he would like
$1 million a month, but will
settle for $8 an hour."

Name _____ Date _____

Practice
For use after Lesson 14.1

Write the ratio as a fraction in simplest form.

1. 8 to 14

2. 36 even : 12 odd

3. 42 vanilla to 48 chocolate

Find the unit rate.

4. $2.50 for 5 ounces

5. 15 degrees in 2 hours

6. 183 miles in 3 hours

Use the ratio table to find the unit rate with the specified units.

7. pounds per box

Boxes	0	1	2	3
Pounds	0	30	60	90

8. cost per notebook

Notebooks	0	5	10	15
Cost (dollars)	0	9.45	18.90	28.35

9. You create 15 centerpieces for a party in 5 hours.

   a. What is the unit rate?

   b. How long will it take you to make 42 centerpieces?

Name_____ Date_____

14.2 Proportions
For use with Activity 14.2

Essential Question How can proportions help you decide when things are "fair"?

---

1 ACTIVITY: Determining Proportions

Work with a partner. Tell whether the two ratios are equivalent. If they are not equivalent, change the next day to make the ratios equivalent. Explain your reasoning.

a. On the first day, you pay $5 for 2 boxes of popcorn. The next day, you pay $7.50 for 3 boxes.

First Day   Next Day

$$\frac{\$5.00}{2 \text{ boxes}} \overset{?}{=} \frac{\$7.50}{3 \text{ boxes}}$$

b. On the first day, it takes you $3\frac{1}{2}$ hours to drive 175 miles.

The next day, it takes you 5 hours to drive 200 miles.

First Day   Next Day

$$\frac{3\frac{1}{2} \text{ h}}{175 \text{ mi}} \overset{?}{=} \frac{5 \text{ h}}{200 \text{ mi}}$$

c. On the first day, you walk 4 miles and burn 300 calories.

The next day, you walk $3\frac{1}{3}$ miles and burn 250 calories.

First Day   Next Day

$$\frac{4 \text{ mi}}{300 \text{ cal}} \overset{?}{=} \frac{3\frac{1}{3} \text{ mi}}{250 \text{ cal}}$$

d. On the first day, you paint 150 square feet in $2\frac{1}{2}$ hours.

The next day, you paint 200 square feet in 4 hours.

First Day   Next Day

$$\frac{150 \text{ ft}^2}{2\frac{1}{2} \text{ h}} \overset{?}{=} \frac{200 \text{ ft}^2}{4 \text{ h}}$$

14.2 **Proportions** (continued)

2 **ACTIVITY:** Checking a Proportion

Work with a partner.

a. It is said that "one year in a dog's life is equivalent to seven years in a human's life." Explain why Newton thinks he has a score of 105 points. Did he solve the proportion correctly?

$$\frac{1 \text{ year}}{7 \text{ years}} \overset{?}{=} \frac{15 \text{ points}}{105 \text{ points}}$$

"I got 15 on my online test. That's 105 in dog points! Isn't that an A+?"

b. If Newton thinks his score is 98 points, how many points does he actually have? Explain your reasoning.

3 **ACTIVITY:** Determining Fairness

Work with a partner. Write a ratio for each sentence. Compare the ratios. If they are equal, then the answer is "It is fair." If they are not equal, then the answer is "It is not fair." Explain your reasoning.

a.

You pay $184 for 2 tickets to a concert. & I pay $266 for 3 tickets to the same concert. ➡ **Is this fair?**

14.2 **Proportions** (continued)

b.

| You get 75 points for answering 15 questions correctly. | **&** | I get 70 points for answering 14 questions correctly. |

 Is this fair?

c.

| You trade 24 football cards for 15 baseball cards. | **&** | I trade 20 football cards for 32 baseball cards. |

Is this fair?

What Is Your Answer?

4. Find a recipe for something you like to eat. Then show how two of the ingredient amounts are proportional when you double or triple the recipe.

5. IN YOUR OWN WORDS How can proportions help you decide when things are "fair"? Give an example.

Name _____ Date _____

14.2 Practice
For use after Lesson 14.2

Tell whether the ratios form a proportion.

1. $\dfrac{1}{5}, \dfrac{5}{15}$

2. $\dfrac{2}{3}, \dfrac{12}{18}$

3. $\dfrac{15}{2}, \dfrac{4}{30}$

4. $\dfrac{56}{21}, \dfrac{8}{3}$

5. $\dfrac{5}{8}, \dfrac{62.5}{100}$

6. $\dfrac{17}{20}, \dfrac{90.1}{106}$

7. $\dfrac{3.2}{4}, \dfrac{16}{24}$

8. $\dfrac{34}{50}, \dfrac{6.8}{10}$

Tell whether the two rates form a proportion.

9. 28 points in 3 games;
   112 points in 12 games

10. 32 notes in 4 measures;
    12 notes in 2 measures

11. You can type 105 words in two minutes. Your friend can type 210 words in four minutes. Are these rates proportional? Explain.

Extension 14.2 Practice
For use after Extension 14.2

Use a graph to tell whether *x* and *y* are in a proportional relationship.

1.

x	2	3	4	5
y	5	7	9	11

2.

x	1	2	3	4
y	3	6	9	12

3.

x	1	2	3	4
y	2.4	4.8	7.2	9.6

4.

x	2	4	6	8
y	2	3	4	5

Extension 14.2 **Practice** (continued)

Interpret each plotted point in the graph of the proportional relationship.

5.

Car

6.

Shrimp

7.

Email

8.

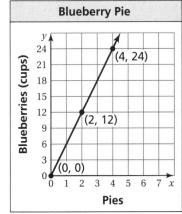

Blueberry Pie

14.3 Writing Proportions
For use with Activity 14.3

Essential Question How can you write a proportion that solves a problem in real life?

---

1 ACTIVITY: Writing Proportions

Work with a partner. A rough rule for finding the correct bat length is "the bat length should be half of the batter's height." So, a 62-inch tall batter uses a bat that is 31 inches long. Write a proportion to find the bat length for each given batter height.

   a. 58 inches   **b.** 60 inches   **c.** 64 inches

---

2 ACTIVITY: Bat Lengths

Work with a partner. Here is a more accurate table for determining the bat length for a batter. Find all the batter heights and corresponding weights for which the rough rule in Activity 1 is exact.

	Height of Batter (inches)							
Weight of Batter (pounds)	45–48	49–52	53–56	57–60	61–64	65–68	69–72	Over 72
Under 61	28	29	29					
61–70	28	29	30	30				
71–80	28	29	30	30	31			
81–90	29	29	30	30	31	32		
91–100	29	30	30	31	31	32		
101–110	29	30	30	31	31	32		
111–120	29	30	30	31	31	32		
121–130	29	30	30	31	32	33	33	
131–140	30	30	31	31	32	33	33	
141–150	30	30	31	31	32	33	33	
151–160	30	31	31	32	32	33	33	33
161–170		31	31	32	32	33	33	34
171–180				32	33	33	34	34
Over 180					33	33	34	34

14.3 Writing Proportions (continued)

3 ACTIVITY: Writing Proportions

Work with a partner. The batting average of a baseball player is the number of "hits" divided by the number of "at bats."

$$\text{batting average} = \frac{\text{hits } (H)}{\text{at Bats } (A)}$$

A player whose batting average is 0.250 is said to be "batting 250."

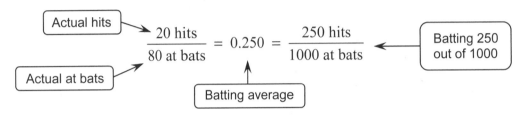

Write a proportion to find how many hits H a player needs to achieve the given batting average. Then solve the proportion.

a. 50 times at bat;
batting average is 0.200.

b. 84 times at bat;
batting average is 0.250.

c. 80 times at bat;
batting average is 0.350.

d. 1 time at bat;
batting average is 1.000.

14.3 **Writing Proportions** (continued)

What Is Your Answer?

4. **IN YOUR OWN WORDS** How can you write a proportion that solves a problem in real life?

5. Two players have the same batting average.

	At Bats	Hits	Batting Average
Player 1	132	45	
Player 2	132	45	

Player 1 gets four hits in the next five at bats. Player 2 gets three hits in the next three at bats.

a. Who has the higher batting average?

b. Does this seem fair? Explain your reasoning.

14.3 Practice
For use after Lesson 14.3

Write a proportion to find how many points a student needs to score on the test to get the given score.

1. test worth 50 points; test score of 84%

2. test worth 75 points; test score of 96%

Use the table to write a proportion.

3.

	Trip 1	Trip 2
Miles	104	78
Gallons	4	g

4.

	Tree 1	Tree 2
Inches	15	x
Years	4	3

Solve the proportion.

5. $\dfrac{1}{3} = \dfrac{x}{12}$

6. $\dfrac{5}{9} = \dfrac{25}{y}$

7. $\dfrac{26}{z} = \dfrac{13}{22}$

8. $\dfrac{b}{30} = \dfrac{2.6}{1.5}$

9. A local Humane Society houses 300 animals. The ratio of cats to all animals is 7 : 15.

   a. Write a proportion that gives the number of cats c.

   b. How many cats are in the Humane Society?

14.4 Solving Proportions
For use with Activity 14.4

Essential Question   How can you use ratio tables and cross products to solve proportions?

---

1   ACTIVITY: Solving a Proportion in Science

Work with a partner. You can use ratio tables to determine the amount of a compound (like salt) that is dissolved in a solution. Determine the unknown quantity. Explain your procedure.

a. **Salt Water**

Salt Water	1 L	3 L
Salt	250 g	x g

I liter   3 liters

There are _____ grams of salt in the 3-liter solution.

b. **White Glue Solution**

Water	$\frac{1}{2}$ cup	1 cup
White Glue	$\frac{1}{2}$ cup	x cups

c. **Borax Solution**

Borax	1 tsp	2 tsp
Water	1 cup	x cups

14.4 **Solving Proportions** (continued)

d. **Slime (See recipe.)**

Borax Solution	$\frac{1}{2}$ cup	1 cup
White Glue Solution	y cups	x cups

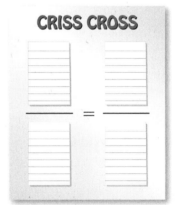

Recipe for SLIME

1. Add ½ cup of water and ½ cup white glue. Mix thoroughly. This is your white glue solution.
2. Add a couple drops of food coloring to the glue solution. Mix thoroughly.
3. Add 1 teaspoon of borax to 1 cup of water. Mix thoroughly. This is your borax solution (about 1 cup).
4. Pour the borax solution and the glue solution into a separate bowl.
5. Place the slime that forms in a plastic bag and squeeze the mixture repeatedly to mix it up.

2 **ACTIVITY:** The Game of Criss Cross

Preparation:

- Cut index cards to make 48 playing cards.

- Write each number on a card.
  1, 1, 1, 2, 2, 2, 3, 3, 3, 4, 4, 4, 5, 5, 5, 6, 6, 6, 7, 7, 7, 8, 8, 8, 9, 9, 9, 10, 10, 10, 12, 12, 12, 13, 13, 13, 14, 14, 14, 15, 15, 15, 16, 16, 16, 18, 20, 25

- Make a copy of the game board.

CRISS CROSS

To play:

- Play with a partner.

- Deal 8 cards to each player.

- Begin by drawing a card from the remaining cards. Use four of your cards to try and form a proportion.

- Lay the four cards on the game board. If you form a proportion, then say "Criss Cross." You earn 4 points. Place the four cards in a discard pile. Now it is your partner's turn.

- If you cannot form a proportion, then it is your partner's turn.

- When the original pile of cards is empty, shuffle the cards in the discard pile. Start again.

- The first player to reach 20 points wins.

Name_____ Date_____

14.4 **Solving Proportions** (continued)

What Is Your Answer?

3. **IN YOUR OWN WORDS** How can you use ratio tables and cross products to solve proportions? Give an example.

4. **PUZZLE** Use each number once to form three proportions.

1	2	10	4	12	20
15	5	16	6	8	3

14.4 Practice

For use after Lesson 14.4

Use multiplication to solve the proportion.

1. $\dfrac{a}{40} = \dfrac{3}{10}$

2. $\dfrac{6}{11} = \dfrac{c}{77}$

3. $\dfrac{b}{65} = \dfrac{7}{13}$

Use the Cross Products Property to solve the proportion.

4. $\dfrac{k}{6} = \dfrac{8}{16}$

5. $\dfrac{5.4}{7} = \dfrac{27}{h}$

6. $\dfrac{8}{11} = \dfrac{4}{y+2}$

Write and solve a proportion to complete the statement.

7. 42 in. = _____ cm

8. 12.6 kg ≈ _____ lb

9. 3 oz ≈ _____ g

10. A cell phone company charges \$5 for 250 text messages. How much does the company charge for 300 text messages?

Name_____ Date_____

Essential Question How can you compare two rates graphically?

┌───┐
1 **ACTIVITY:** Comparing Unit Rates
└───┘

Work with a partner. The table shows the maximum speeds of several animals.

   a. Find the missing speeds. Round your answers to the nearest tenth.

   b. Which animal is fastest? Which animal is slowest?

   c. Explain how you convert between the two units of speed.

Animal	Speed (miles per hour)	Speed (feet per second)
Antelope	61.0	
Black mamba snake		29.3
Cheetah		102.6
Chicken		13.2
Coyote	43.0	
Domestic pig		16.0
Elephant		36.6
Elk		66.0
Giant tortoise	0.2	
Giraffe	32.0	
Gray fox		61.6
Greyhound	39.4	
Grizzly bear		44.0
Human		41.0
Hyena	40.0	
Jackal	35.0	
Lion		73.3
Peregrine falcon	200.0	
Quarter horse	47.5	
Spider		1.76
Squirrel	12.0	
Thomson's gazelle	50.0	
Three-Toed sloth		0.2
Tuna	47.0	

14.5 **Slope** (continued)

2 **ACTIVITY:** Comparing Two Rates Graphically

Work with a partner. A cheetah and a Thomson's gazelle run at maximum speeds.

    a. Use the table in Activity 1 to calculate the missing distances.

	Cheetah	Gazelle
Time (seconds)	**Distance (feet)**	**Distance (feet)**
0		
1		
2		
3		
4		
5		
6		
7		

    b. Use the table to write ordered pairs. Then plot the ordered pairs and connect the points for each animal. What do you notice about the graphs?

    c. Which graph is steeper? The speed of which animal is greater?

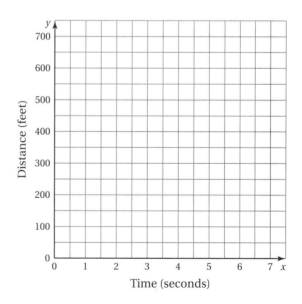

14.5 Slope (continued)

What Is Your Answer?

3. IN YOUR OWN WORDS How can you compare two rates graphically?
Explain your reasoning. Give some examples with your answer.

4. REPEATED REASONING Choose 10 animals from Activity 1.

   a. Make a table for each animal similar to the table in Activity 2.

   b. Sketch a graph of the distances for each animal.

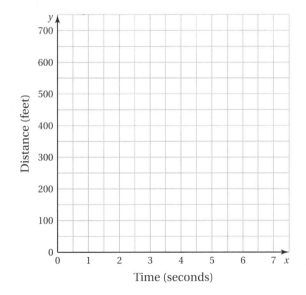

   c. Compare the steepness of the 10 graphs. What can you conclude?

Name_____ Date_____

Find the slope of the line.

1.

2.

3.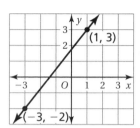

Graph the data. Then find and interpret the slope of the line through the points.

4.

Minutes, x	0	1	3	5
Pages, y	0	1.5	4.5	7.5

5.

Miles, x	0	1	2	3
Calories, y	0	135	270	405

6. By law, the maximum slope of a wheelchair ramp is $\frac{1}{12}$.

   a. A ramp is designed that is 4 feet high and has a horizontal length of 50 feet. Does this ramp meet the law? Explain.

   b. What could be adjusted on an unacceptable ramp so that it meets the law?

14.6 Direct Variation
For use with Activity 14.6

Essential Question How can you use a graph to show the relationship between two quantities that vary directly? How can you use an equation?

1 ACTIVITY: Math in Literature

Gulliver's Travels was written by Jonathan Swift and published in 1726. Gulliver was shipwrecked on an island in Lilliput, where the people were only 6 inches tall. When the Lilliputians decided to make a shirt for Gulliver, a Lilliputian tailor stated that he could determine Gulliver's measurements by simply measuring the distance around Gulliver's thumb. He said "Twice around the thumb equals once around the wrist. Twice around the wrist is once around the neck. Twice around the neck is once around the waist."

Work with a partner. Use the tailor's statement to complete the table.

Thumb, *t*	Wrist, *w*	Neck, *n*	Waist, *x*
0 in.			
1 in.			
	4 in.		
		12 in.	
			32 in.
	10 in.		

14.6 **Direct Variation** (continued)

2 **ACTIVITY:** Drawing a Graph

Work with a partner. Use the information from Activity 1.

    a. In your own words, describe the relationship between *t* and *w*.

    b. Use the table to write ordered pairs (*t*, *w*). Then plot the ordered pairs.

    c. What do you notice about the graph of the ordered pairs?

    d. Choose two points and find the slope of the line between them.

    e. The quantities *t* and *w* are said to *vary directly*. An equation that describes the relationship is *w* = _____ *t*.

3 **ACTIVITY:** Drawing a Graph and Writing an Equation

Work with a partner. Use the information from Activity 1 to draw a graph of the relationship. Write an equation that describes the relationship between the two quantities.

    a. Thumb *t* and neck *n*

$$\left(n = \boxed{} \; t \right)$$

    b. Wrist *w* and waist *x*

$$\left(x = \boxed{} \; w \right)$$

14.6 **Direct Variation** (continued)

c. Wrist w and thumb t

$$\left(t = \boxed{} \, w \right)$$

d. Waist x and wrist w

$$\left(w = \boxed{} \, x \right)$$

What Is Your Answer?

4. IN YOUR OWN WORDS How can you use a graph to show the relationship between two quantities that vary directly? How can you use an equation?

5. STRUCTURE How are all the graphs in Activity 3 alike?

6. Give a real-life example of two variables that vary directly.

7. Work with a partner. Use string to find the distance around your thumb, wrist, and neck. Do your measurements agree with the tailor's statements in *Gulliver's Travels*? Explain your reasoning.

Name _____ Date _____

Tell whether x and y show direct variation. Explain your reasoning. If so, find k.

1.

x	1	2	3	4
y	3	6	9	12

2.

x	-1	0	1	2
y	1	3	7	13

3.

x	0	2	4	6
y	8	5	2	-1

4. $y + 2 = x$

5. $3y = x$

6. $\dfrac{y}{x} = 4$

The variables x and y vary directly. Use the values to find the constant of proportionality and write an equation that relates x and y.

7. $y = 8; x = 2$

8. $y = 14, x = 16$

9. $y = 25, x = 35$

10. The table shows the cups c of dog food needed to feed a dog that weighs p pounds. Graph the data. Tell whether p and c show direct variation. If so, write an equation that represents the line.

Pounds, p	10	20	40	70
Food, c	$\dfrac{3}{4}$	$1\dfrac{1}{4}$	2	$2\dfrac{3}{4}$

Chapter 15 Fair Game Review

Write the percent as a fraction or mixed number in simplest form.

1. 25%

2. 65%

3. 110%

4. 250%

5. 15%

6. 6%

7. A store marks up a pair of sneakers 30%. Write the percent as a fraction or mixed number in simplest form.

Big Ideas Math Advanced 1 **331**
Record and Practice Journal

Chapter
15 **Fair Game Review** (continued)

Write the fraction or mixed number as a percent.

8. $\dfrac{1}{5}$

9. $\dfrac{1}{4}$

10. $\dfrac{21}{25}$

11. $1\dfrac{2}{5}$

12. $2\dfrac{13}{20}$

13. $1\dfrac{1}{2}$

14. You own $\dfrac{3}{5}$ of the coins in a collection. What percent of the coins do you own?

15.1 Percents and Decimals
For use with Activity 15.1

Essential Question How does the decimal point move when you rewrite a percent as a decimal and when you rewrite a decimal as a percent?

---

1 **ACTIVITY:** Writing Percents as Decimals

Work with a partner. Write the percent shown by the model. Write the percent as a decimal.

a.

$$\boxed{} = \frac{\boxed{}}{\boxed{}} \quad \leftarrow \text{per} \quad \leftarrow \text{cent}$$

$$= \frac{\boxed{}}{\boxed{}} \quad \text{Simplify.}$$

$$= \boxed{} \quad \text{Write fraction as a decimal.}$$

b.

c.

d.

e.

15.1 **Percents and Decimals** (continued)

f.

g.

2 **ACTIVITY:** Writing Percents as Decimals

Work with a partner. Write the percent as a decimal.

a. $13.5\% = \dfrac{\boxed{}}{\boxed{}}$ per cent

$= \dfrac{\boxed{}}{\boxed{}}$ Multiply numerator and denominator by 10.

$= \boxed{}$ Write fraction as a decimal.

b. 12.5% **c.** 3.8% **d.** 0.5%

15.1 Percents and Decimals (continued)

3 **ACTIVITY:** Writing Decimals as Percents

Work with a partner. Draw a model to represent the decimal. Write the decimal as a percent.

    a. 0.1   **b.** 0.24

    c. 0.58   **d.** 0.05

What Is Your Answer?

4. **IN YOUR OWN WORDS** How does the decimal point move when you rewrite a percent as a decimal and when you rewrite a decimal as a percent?

5. Explain why the decimal point moves when you rewrite a percent as a decimal and when you rewrite a decimal as a percent.

Name _____ Date _____

15.1 Practice
For use after Lesson 15.1

Write the percent as a decimal.

1. 35%

2. 160%

3. 74.8%

4. 0.3%

Write the decimal as a percent.

5. 1.23

6. 0.49

7. 0.024

8. 0.881

Write the percent as a fraction in simplest form and as a decimal.

9. 48%

10. 15.5%

11. 84.95%

12. People with severe hearing loss were given a sentence and word recognition test six months after they got implants in their ears. The patients scored an average of 82% on the test. Write this percent as a decimal.

Name_____ Date _____

Essential Question How can you order numbers that are written as fractions, decimals, and percents?

---

1 **ACTIVITY:** Using Fractions, Decimals, and Percents

Work with a partner. Decide which number form (fraction, decimal, or percent) is more common. Then find which is greater.

a. 7% sales tax or $\dfrac{1}{20}$ sales tax

b. 0.37 cup of flour or $\dfrac{1}{3}$ cup of flour

c. $\dfrac{5}{8}$-inch wrench or 0.375-inch wrench

d. $12\dfrac{3}{5}$ dollars or 12.56 dollars

e. 93% test score or $\dfrac{7}{8}$ test score

f. $5\dfrac{5}{6}$ fluid ounces or 5.6 fluid ounces

---

2 **ACTIVITY:** Ordering Numbers

Work with a partner to order the following numbers.

$\dfrac{1}{8}$ 11% $\dfrac{3}{20}$ 0.172   0.32   43%   7%   0.7 $\dfrac{5}{6}$

a. Decide on a strategy for ordering the numbers. Will you write them all as fractions, decimals, or percents?

b. Use your strategy and a number line to order the numbers from least to greatest. (Note: Label the number line appropriately.)

15.2 Comparing and Ordering Fractions, Decimals, and Percents (continued)

3 **ACTIVITY:** The Game of Math Card War

Preparation:

- Cut index cards to make 40 playing cards.*

- Write each number in the table onto a card.

75%	$\frac{3}{4}$	$\frac{1}{3}$	$\frac{3}{10}$	0.3	25%	0.4	0.25	100%	0.27
0.75	$66\frac{2}{3}\%$	12.5%	40%	$\frac{1}{4}$	4%	0.5%	0.04	$\frac{1}{100}$	$\frac{2}{3}$
0	30%	5%	$\frac{27}{100}$	0.05	$33\frac{1}{3}\%$	$\frac{2}{5}$	0.333...	27%	1%
1	0.01	$\frac{1}{20}$	$\frac{1}{8}$	0.125	$\frac{1}{25}$	$\frac{1}{200}$	0.005	0.666...	0%

To Play:

- Play with a partner.

- Deal 20 cards to each player facedown.

- Each player turns one card faceup. The player with the greater number wins. The winner collects both cards and places them at the bottom of his or her cards.

- Suppose there is a tie. Each player lays three cards facedown, then a new card faceup. The player with the greater of these new cards wins. The winner collects all 10 cards and places them at the bottom of his or her cards.

- Continue playing until one player has all the cards. This player wins the game.

*Cut-outs are available in the back of the Record and Practice Journal.

15.2 Comparing and Ordering Fractions, Decimals, and Percents (continued)

What Is Your Answer?

4. **IN YOUR OWN WORDS** How can you order numbers that are written as fractions, decimals, and percents? Give an example with your answer.

5. All but one of the U.S. coins shown has a name that is related to its value. Which one is it? How are the names of the others related to their values?

Name_____   Date _____

15.2 **Practice**
          For use after Lesson 15.2

Circle the number that is greater.

1. 0.06, 60% **2.** 78%, $\dfrac{19}{25}$ **3.** $\dfrac{23}{20}$, 110% **4.** 0.23, 2.3%

Use a number line to order the numbers from least to greatest.

5. 44.5%, 0.4445, $\dfrac{4}{9}$, 0.44

6. $\dfrac{5}{12}$, 0.4, 42%, 0.416

7. The table shows the portion of each age group that recycles plastic. Order the groups by the portion that recycle from least to greatest.

Age Group	Echo Boomers	Gen X	Baby Boomers	Matures
Portion that Recycle	51%	0.57	0.61	$\dfrac{6}{10}$

15.3 The Percent Proportion
For use with Activity 15.3

Essential Question How can you use models to estimate percent questions?

---

1 **ACTIVITY:** Estimating a Part

The statement "25% of 12 is 3" has three numbers. In real-life problems, any one of these can be unknown.

Question	Which number is missing?	Type of Question
What is 25% of 12?	_____	Find a part of a number.
3 is what percent of 12?	_____	Find a percent.
3 is 25% of what?	_____	Find the whole.

Work with a partner. Use a model to estimate the answer to each question.

a. What number is 50% of 30?

```
0%                                          100%
┌──────────────────────────────────────────────┐
│                                                │
└──────────────────────────────────────────────┘
0                                           30
```

b. What number is 75% of 30?

```
0%                    100%
┌──────────────────────┐
│                      │
└──────────────────────┘
0                     30
```

c. What number is 40% of 30?

d. What number is 6% of 30?

e. What number is 65% of 30?

```
0%                    100%
┌──────────────────────┐
│                      │
└──────────────────────┘
0                     30
```

Name _____ Date _____

15.3 **The Percent Proportion** (continued)

---

2 ACTIVITY: Estimating a Percent

Work with a partner. Use a model to estimate the answer to each question.

  a. 15 is what percent of 75?

0%   100%

0   75

  b. 5 is what percent of 20?

0%   100%

0   20

  c. 18 is what percent of 40?

0%   100%

0   40

  d. 50 is what percent of 80?

0%   100%

0   80

  e. 75 is what percent of 50?

0%   100%

0   50

---

3 ACTIVITY: Estimating a Whole

Work with a partner. Use a model to estimate the answer to each question.

  a. 24 is $33\frac{1}{3}\%$ of what number?

0%   100%

  b. 13 is 25% of what number?

0%   100%

  c. 110 is 20% of what number?

0%   100%

15.3 **The Percent Proportion** (continued)

d. 75 is 75% of what number?   **e.** 81 is 45% of what number?

0%   100%   0%   100%

---

4 **ACTIVITY:** Using Ratio Tables

Work with a partner. Use a ratio table to answer each question. Then compare your answer to the estimate you found using the model.

1d a. What number is 6% of 30?   **1e b.** What number is 65% of 30?

Part	6		
Whole	100		30

Part	65		
Whole	100		30

2c c. 18 is what percent of 40?   **3e d.** 81 is 45% of what number?

Part	18		
Whole	40		100

Part	45		81
Whole	100		

What Is Your Answer?

5. IN YOUR OWN WORDS How can you use models to estimate percent questions? Give examples to support your answer.

6. Complete the proportion below using the given labels.

percent	whole

100	part

$$\frac{}{} = \frac{}{}$$

15.3 Practice
For use after Lesson 15.3

Write and solve a proportion to answer the question.

1. 40% of 60 is what number?

2. 17 is what percent of 50?

3. 38% of what number is 57?

4. 44% of 25 is what number?

5. 52 is what percent of 50?

6. 150% of what number is 18?

7. You put 60% of your paycheck into your savings account. Your paycheck is $235. How much money do you put in your savings account?

15.4 The Percent Equation
For use with Activity 15.4

Essential Question How can you use an equivalent form of the percent proportion to solve a percent problem?

---

1 ACTIVITY: Solving Percent Problems Using Different Methods

Work with a partner. The circle graph shows the number of votes received by each candidate during a school election. So far, only half the students have voted.

Votes Received by Each Candidate

a. Complete the table.

b. Find the percent of students who voted for each candidate. Explain the method you used to find your answers.

Candidate	Number of votes received / Total number of votes
Sue	
Miguel	
Leon	
Hong	

c. Compare the method you used in part (b) with the methods used by other students in your class. Which method do you prefer? Explain.

15.4 **The Percent Equation** (continued)

2 **ACTIVITY:** Finding Parts Using Different Methods

Work with a partner. The circle graph shows the final results of the election.

a. Find the number of students who voted for each candidate. Explain the method you used to find your answers.

Final Results

Hong 25%

Sue 20%

Miguel 20%

Leon 35%

b. Compare the method you used in part (a) with the methods used by other students in your class. Which method do you prefer? Explain.

3 **ACTIVITY:** Deriving the Percent Equation

Work with a partner. In Section 15.3, you used the percent proportion to find the missing percent, part, or whole. You can also use the *percent equation* to find these missing values.

a. Complete the steps below to find the percent equation.

$$\frac{\text{part}}{\text{whole}} = \text{percent}$$ Definition of percent

$$\frac{\text{part}}{\text{whole}} \cdot \boxed{} = \boxed{} \cdot \boxed{}$$ Multiply each side by the $\boxed{}$.

$$\text{part} = \boxed{} \cdot \boxed{}$$ Divide out common factors. This is the percent equation.

b. You used two methods in Activity 2 to find the number of students who voted for each candidate. Do you prefer the percent proportion or the percent equation method?

15.4 **The Percent Equation** (continued)

4 **ACTIVITY:** Identifying Different Equations

Work with a partner. Without doing any calculations, choose the equation that you cannot use to answer each question.

a. What number is 55% of 80?

| $a = 0.55 \cdot 80$ | $a = \dfrac{11}{20} \cdot 80$ | $80a = 0.55$ | $\dfrac{a}{80} = \dfrac{55}{100}$ |

b. 24 is 60% of what number?

| $\dfrac{24}{w} = \dfrac{60}{100}$ | $24 = 0.6 \cdot w$ | $\dfrac{24}{60} = w$ | $24 = \dfrac{3}{5} \cdot w$ |

What Is Your Answer?

5. IN YOUR OWN WORDS How can you use an equivalent form of the percent proportion to solve a percent problem?

6. Write a percent proportion and a percent equation that you can use to answer the question below.

16 is what percent of 250?

15.4 Practice
For use after Lesson 15.4

Write and solve an equation to answer the question.

1. What number is 35% of 80?

2. 8 is what percent of 5?

3. What percent of 125 is 50?

4. 12% of what number is 48?

5. 12 is what percent of 50?

6. What percent of 12 is 3?

7. You receive 15% of the profit from a car wash. How much money do you receive from a profit of $300?

15.5 Percents of Increase and Decrease
For use with Activity 15.5

Essential Question What is a percent of decrease? What is a percent of increase?

---

1 **ACTIVITY:** Percent of Decrease

Work with a partner.

Each year in the Columbia River Basin, adult salmon swim upriver to streams to lay eggs and hatch their young.

To go up river, the adult salmon use fish ladders. But to go down the river, the young salmon must pass through several dams.

At one time, there were electric turbines at each of the eight dams on the main stem of the Columbia and Snake Rivers. About 88% of the young salmon passed through these turbines unharmed.

a. Complete the table to show the number of young salmon that made it through the dams.

Dam	0	1	2	3	4	5	6	7	8
Salmon	1000								

15.5 **Percents of Increase and Decrease** (continued)

b. Display the data in a bar graph.

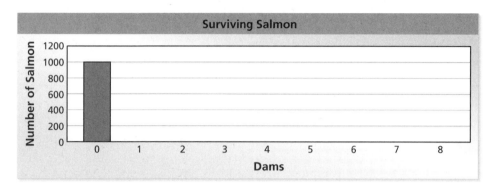

c. By what percent did the number of young salmon decrease when passing through each dam?

2 **ACTIVITY:** Percent of Increase

Work with a partner. In 2013, the population of a city was 18,000 people.

a. An organization projects that the population will increase by 2% each year for the next 7 years. Complete the table to find the populations of the city for 2014 through 2020. Then display the data in a bar graph.

For 2014: 2% of 18,000 $= 0.02 \bullet 18,000 = 360$

$18,000 + 360 = 18,360$ ← 2014 Population

↑ 2013 Population   ↑ Increase

Year	Population
2013	18,000
2014	18,360
2015	
2016	
2017	
2018	
2019	
2020	

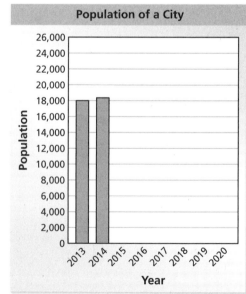

15.5 **Percents of Increase and Decrease** (continued)

b. Another organization projects that the population will increase by 3% each
year for the next 7 years. Repeat part (a) using this percent.

Year	Population
2013	
2014	
2015	
2016	
2017	
2018	
2019	
2020	

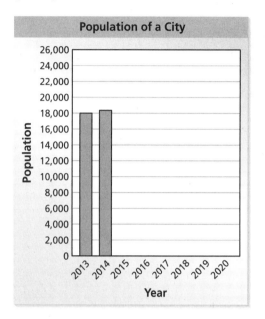

c. Which organization projects the larger populations? How many more
people do they project for 2020?

What Is Your Answer?

3. **IN YOUR OWN WORDS** What is a percent of decrease? What is a percent
of increase?

4. Describe real-life examples of a percent of decrease and a percent of increase.

Name _____ Date _____

15.5 Practice
For use after Lesson 15.5

Find the new amount.

1. 120 books increased by 55%

2. 80 members decreased by 65%

Identify the percent of change as an *increase* or *decrease*. Then find the percent of change. Round to the nearest tenth of a percent, if necessary.

3. 25 points to 50 points

4. 125 invitations to 75 invitations

5. 32 pages to 28 pages

6. 7 players to 10 players

7. One week, 72 people got a speeding ticket. The next week, only 36 people got a speeding ticket. What is the percent of change in speeding tickets?

15.6 Discounts and Markups

For use with Activity 15.6

Essential Question How can you find discounts and selling prices?

---

1 ACTIVITY: Comparing Discounts

Work with a partner. The same pair of sneakers is on sale at three stores. Which one is the best buy? Explain.

a. Regular Price: $45   **b.** Regular Price: $49   **c.** Regular Price: $39

a.

$0 $9 $18 $27 $36 $45

b.

$0 $9.80 $19.60 $29.40 $39.20 $49

c.

$0 $7.80 $15.60 $23.40 $31.20 $39

15.6 Discounts and Markups (continued)

2 ACTIVITY: Finding the Original Price

Work with a partner.

a. You buy a shirt that is on sale for 30% off. You pay $22.40. Your friend wants to know the original price of the shirt. Show how you can use the model to find the original price.

$0 $22.40   Original
                                                      Price

b. Explain how you can use the percent proportion to find the original price.

3 ACTIVITY: Finding Selling Prices

You own a small jewelry store. You increase the price of jewelry by 125%.

Work with a partner. Use a model to estimate the selling price of the jewelry. Then use a calculator to find the selling price.

a. Your cost is $250.

$0 $250   Selling
                                                      Price

15.6 **Discounts and Markups** (continued)

b. Your cost is $50.

c. Your cost is $170.

What Is Your Answer?

4. **IN YOUR OWN WORDS** How can you find discounts and selling prices?
   Give examples of each.

15.6 Practice
For use after Lesson 15.6

Complete the table.

	Original Price	Percent of Discount	Sale Price
1.	$20	20%	
2.	$95	35%	
3.		75%	$55.50
4.		40%	$78

Find the selling price.

5. Cost to store: $20

Markup: 15%

6. Cost to store: $56

Markup: 80%

7. Cost to store: $110

Markup: 140%

8. A store buys an item for $10. To earn a profit of $25, what percent does the store need to markup the item?

Name_____ Date_____

15.7 Simple Interest
For use with Activity 15.7

Essential Question How can you find the amount of simple interest earned on a savings account? How can you find the amount of interest owed on a loan?

Simple interest is money earned on a savings account or an investment. It can also be money you pay for borrowing money.

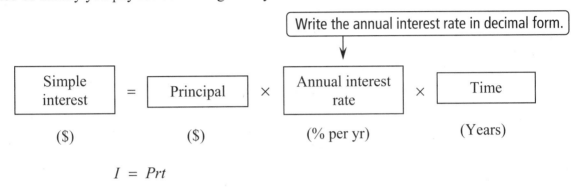

Write the annual interest rate in decimal form.

Simple interest	=	Principal	×	Annual interest rate	×	Time
($)		($)		(% per yr)		(Years)

$$I = Prt$$

---

1 ACTIVITY: Finding Simple Interest

Work with a partner. You put $100 in a savings account. The account earns 6% simple interest per year. (a) Find the interest earned and the balance at the end of 6 months. (b) Complete the table. Then make a bar graph that shows how the balance grows in 6 months.

a. $I = Prt$

b.

Time	Interest	Balance
0 month		
1 month		
2 months		
3 months		
4 months		
5 months		
6 months		

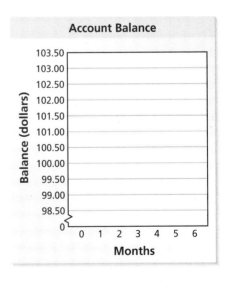

Account Balance

15.7 **Simple Interest** (continued)

2 **ACTIVITY:** Financial Literacy

Work with a partner. Use the following information to write a report about credit cards. In the report, describe how a credit card works. Include examples that show the amount of interest paid each month on a credit card.

U.S. Credit Card Data

- A typical household with credit card debt in the United States owes about $16,000 to credit card companies.

- A typical credit card interest rate is 14% to 16% per year. This is called the annual percentage rate.

3 **ACTIVITY:** The National Debt

Work with a partner. In 2012, the United States owed about $16 trillion in debt. The interest rate on the national debt is about 1% per year.

a. Write $16 trillion in decimal form. How many zeros does this number have?

15.7 Simple Interest (continued)

b. How much interest does the United States pay each year on its national debt?

c. How much interest does the United States pay each day on its national debt?

d. The United States has a population of about 314 million people. Estimate the amount of interest that each person pays per year toward interest on the national debt.

What Is Your Answer?

4. **IN YOUR OWN WORDS** How can you find the amount of simple interest earned on a savings account? How can you find the amount of interest owed on a loan? Give examples with your answer.

15.7 Practice
For use after Lesson 15.7

An account earns simple interest. (a) Find the interest earned. (b) Find the balance of the account.

1. $400 at 7% for 3 years

2. $1200 at 5.6% for 4 years

Find the annual interest rate.

3. $I = \$18$, $P = \$200$, $t = 18$ months

4. $I = \$310$, $P = \$1000$, $t = 5$ years

Find the amount of time.

5. $I = \$60$, $P = \$750$, $r = 4\%$

6. $I = \$825$, $P = \$2500$, $r = 5.5\%$

7. You put $500 in a savings account. The account earns $15.75 simple interest in 6 months. What is the annual interest rate?

Glossary

This student friendly glossary is designed to be a reference for key vocabulary, properties, and mathematical terms. Several of the entries include a short example to aid your understanding of important concepts.

Also available at *BigIdeasMath.com:*

- multi-language glossary
- vocabulary flash cards

absolute value The distance between a number and 0 on a number line; The absolute value of a number a is written as $\lvert a \rvert$. $$\lvert -5 \rvert = 5$$ $$\lvert 5 \rvert = 5$$	**addend** A number to be added to another number 2 or 3 in the sum $2 + 3$
Addition Property of Equality When you add the same number to each side of an equation, the two sides remain equal. $$\begin{array}{rcl} x - 4 &=& 5 \\ +\,4 & & +\,4 \\ \hline x &=& 9 \end{array}$$	**Addition Property of Inequality** When you add the same number to each side of an inequality, the inequality remains true. $$\begin{array}{rcl} x - 4 &>& 5 \\ +\,4 & & +\,4 \\ \hline x &>& 9 \end{array}$$
Addition Property of Zero The sum of any number and 0 is that number. $$5 + 0 = 5$$	**additive inverse** The opposite of a number The additive inverse of 8 is -8.

Additive Inverse Property	**algebraic expression**
The sum of an integer and its additive inverse is 0. $$8 + (-8) = 0$$	An expression that contains numbers, operations, and one or more symbols $$8 + x, 6 \times a - b$$
angle	**area**
A figure formed by two rays with the same endpoint	The amount of surface covered by a figure; Area is measured in square units such as square feet (ft^2) or square meters (m^2). 5 units / 3 units $$A = 5 \times 3 = 15 \text{ square units}$$
Associative Properties of Addition and Multiplication	**bar graph**
Changing the grouping of addends or factors does not change the sum or product. $$(3 + 4) + 5 = 3 + (4 + 5)$$ $$(3 \bullet 4) \bullet 5 = 3 \bullet (4 \bullet 5)$$	A graph in which the lengths of bars are used to represent and compare data
base (of a power)	**box-and-whisker plot**
The base of a power is the repeated factor. *See power.*	A type of graph that represents a data set along a number line by using the least value, the greatest value, and the quartiles of the data

capacity	**coefficient**
The amount a container can hold	The numerical factor of a term that contains a variable In the algebraic expression $6k + 8$, 6 is the coefficient of the term $6k$.

common factors	**common multiples**
Factors that are shared by two or more numbers 2 is a common factor of 8 and 10.	Multiples that are shared by two or more numbers Multiples of 4: 4, 8, 12, 16, 20, 24, ... Multiples of 6: 6, 12, 18, 24, 30, 36, ... The first two common multiples of 4 and 6 are 12 and 24.

Commutative Properties of Addition and Multiplication	**complex fraction**
Changing the order of addends or factors does not change the sum or product. $$2 + 8 = 8 + 2$$ $$2 \bullet 8 = 8 \bullet 2$$	A fraction that has at least one fraction in the numerator, denominator, or both $$\dfrac{\frac{1}{4}}{\frac{1}{2}}$$

composite figure	**composite number**
A figure made up of triangles, squares, rectangles, and other two-dimensional figures 	A whole number greater than 1 that has factors other than 1 and itself 4, 6, 8, 9, 10, 12, 14, 15, 16, 18, 20, ...

constant	**constant of proportionality**
A term without a variable	The number k in the direct variation equation $y = kx$
In the expression $2x + 8$, the term 8 is a constant.	The constant of proportionality in the equation $y = 2x$ is 2.

conversion factor	**coordinate plane**
A rate that equals 1; A conversion factor is used to convert units.	A coordinate plane is formed by the intersection of a horizontal number line and a vertical number line.
1 mile = 5280 feet	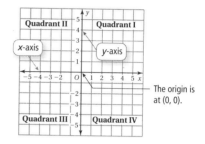

cross products	**Cross Products Property**
In the proportion $\dfrac{a}{b} = \dfrac{c}{d}$, the products $a \bullet d$ and $b \bullet c$ are called cross products.	The cross products of a proportion are equal.
$2 \bullet 6$ and $3 \bullet 4$	$2 \bullet 6 = 3 \bullet 4$

cubic units	**dependent variable**
The units volume is measured in	The variable whose value depends on the independent variable in an equation in two variables
cubic feet $\left(\text{ft}^3\right)$, cubic meters $\left(\text{m}^3\right)$.	In the equation $y = 5x - 8$, y is the dependent variable.

difference

The result when one number is subtracted from another number

The difference of 4 and 3 is $4 - 3$, or 1.

direct variation

Two quantities x and y show direct variation when $y = kx$, where k is a number and $k \neq 0$.

The graph of $y = kx$ is a line with a slope of k that passes through the origin.

discount

A decrease in the original price of an item

The original price of a pair of shoes is \$95. The sale price is \$65. The discount is \$30.

Distributive Property

To multiply a sum or difference by a number, multiply each number in the sum or difference by the number outside the parentheses. Then evaluate.

$$3(12 + 9) = 3(12) + 3(9)$$
$$3(12 - 9) = 3(12) - 3(9)$$

Division Property of Equality

When you divide each side of an equation by the same nonzero number, the two sides remain equal.

$$4x = 32$$
$$\frac{4x}{4} = \frac{32}{4}$$
$$x = 8$$

Division Property of Inequality

When you divide each side of an inequality by the same positive number, the inequality remains true.

$$4x < 8$$
$$\frac{4x}{4} < \frac{8}{4}$$
$$x < 2$$

edge

A line segment where two faces intersect

See face.

equation

A mathematical sentence that uses an equal sign, =, to show that two expressions are equal

$$4x = 16, \ a + 7 = 21$$

equation in two variables An equation that represents two quantities that change in relationship to one another $$y = 2x, y = 4x - 3$$	**equivalent equations** Equations that have the same solutions $$2x - 8 = 0 \text{ and } 2x = 8$$
equivalent expressions Expressions with the same value $$7 + 4, 4 + 7$$	**equivalent rates** Rates that have the same unit rate 6 miles in 3 hours and 4 miles in 2 hours
equivalent ratios Two ratios that describe the same relationship $$2 : 3 \text{ and } 4 : 6$$	**evaluate (a numerical expression)** Use the order of operations to find the value of a numerical expression. *See order of operations.*
exponent The exponent of a power indicates the number of times the base is used as a factor. *See power.*	**expression** A mathematical phrase containing numbers, operations, and/or variables *See numerical expression or algebraic expression.*

face	**factor**
A flat surface of a polyhedron 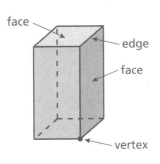	When whole numbers other than zero are multiplied together, each number is a factor of the product. $2 \times 3 \times 4 = 24$, so 2, 3, and 4 are factors of 24.
factor pair Two whole numbers other than zero that are multiplied together to get a product Because $2 \bullet 5 = 10$, the pair 2, 5 is a factor pair of 10.	**factor tree** A diagram that shows the prime factorization of a number $60 = 2 \bullet 2 \bullet 3 \bullet 5$, or $2^2 \bullet 3 \bullet 5$
factoring an expression Writing a numerical expression or algebraic expression as a product of factors $5x - 15 = 5(x - 3)$	**first quartile** (Q_1) The median of the lower half of a data set *See quartiles.*
five-number summary The five numbers that make up a box-and-whisker plot least value, first quartile, median, third quartile, greatest value	**frequency** The number of data values in an interval *See frequency table or histogram.*

frequency table

A table used to group data values into intervals

Pairs of Shoes	Frequency
1–5	11
6–10	4
11–15	0
16–20	3
21–25	6

graph of an inequality

A graph that shows all the solutions of an inequality on a number line

$x > 2$

greatest common factor (GCF)

The greatest of the common factors of two or more numbers

The common factors of 12 and 20 are 1, 2, and 4. So the GCF of 12 and 20 is 4.

histogram

A bar graph that shows the frequency of data values in intervals of the same size; The height of a bar represents the frequency of the values in the interval. There are no spaces between bars.

independent variable

The variable representing the quantity that can change freely in an equation in two variables

In the equation $y = 5x - 8$, x is the independent variable.

inequality

A mathematical sentence that compares expressions; It contains the symbols <, >, ≤, or ≥.

$x - 4 < 14,\ x + 5 \geq 67$

integers

The set of whole numbers and their opposites

..., −3, −2, −1, 0, 1, 2, 3, ...

interest

Money paid or earned for the use of money

See simple interest.

interquartile range (IQR) The difference between the third quartile and the first quartile of a data set; represents the range of the middle half of the data The interquartile range of the data set 3, 4, 8, 16, 21, 26 is $21 - 4 = 17$.	**inverse operations** Operations that "undo" each other, such as addition and subtraction or multiplication and division
leaf Digit or digits on the right of a stem-and-leaf plot *See stem-and-leaf plot.*	**least common denominator (LCD)** The least common multiple of the denominators of two or more fractions The least common denominator of $\frac{3}{4}$ and $\frac{5}{6}$ is the least common multiple of 4 and 6, or 12.
least common multiple (LCM) The least of the common multiples of two or more numbers Multiples of 10: 10, 20, 30, 40, … Multiples of 15: 15, 30, 45, 60, … The least common multiple of 10 and 15 is 30.	**like terms** Terms of an algebraic expression that have the same variables raised to the same exponents 4 and 8, $2x$ and $7x$
line segment Part of a line that consists of two points, called endpoints, and all the points on the line between the endpoints 	**linear expression** An algebraic expression in which the exponent of the variable is 1 $-4x,\ 3x + 5,\ 5 - \frac{1}{6}x$

markup The increase from what a store pays to the selling price A store buys a hat for $12 and sells it for $20. The markup is $8.	**mean** The sum of the data divided by the number of data values The mean of the values 7, 4, 8, and 9 is $$\frac{7 + 4 + 8 + 9}{4} = \frac{28}{4} = 7.$$
mean absolute deviation An average of how much data values differ from the mean The mean of the data set 5, 7, 12, 16 is 10. The sum of the distances between each data value and the mean is 16. So, the mean absolute deviation is $\frac{16}{4} = 4.$	**measure of center** A measure that describes the typical value of a data set The mean, median, and mode are all measures of center.
measure of variation A measure that describes the distribution of a data set The range, interquartile range, and mean absolute deviation are all measures of variation.	**median** For a data set with an odd number of ordered values, the median is the middle value. For a data set with an even number of ordered values, the median is the mean of the two middle values. The median of the data set 24, 25, 29, 33, 38 is 29 because 29 is the middle value.
metric system Decimal system of measurement, based on powers of 10, that contains units for length, capacity, and mass centimeter, meter, liter, kilogram	**mode** The data value or values that occur most often; Data can have one mode, more than one mode, or no mode. The modes of the data set 3, 4, 4, 7, 7, 9, 12 are 4 and 7 because they occur most often.

Multiplication Properties of Zero and One

The product of any number and 0 is 0.

The product of any number and 1 is that number.

$$5 \bullet 0 = 0$$
$$6 \bullet 1 = 6$$

Multiplication Property of Equality

When you multiply each side of an equation by the same nonzero number, the two sides remain equal.

$$\frac{x}{4} = 2$$
$$\frac{x}{4} \bullet 4 = 2 \bullet 4$$
$$x = 8$$

Multiplication Property of Inequality

When you multiply each side of an inequality by the same positive number, the inequality remains true.

$$\frac{x}{4} < 2$$
$$\frac{x}{4} \bullet 4 < 2 \bullet 4$$
$$x < 8$$

Multiplicative Inverse Property

The product of a nonzero number and its reciprocal is 1.

$$5 \bullet \frac{1}{5} = 1$$

negative numbers

Numbers that are less than 0

$$-10, -500, -10{,}000$$

net

A two-dimensional representation of a solid

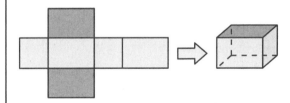

number line

A line whose points are associated with numbers that increase from left to right

numerical expression

An expression that contains only numbers and operations

$$12 + 6, 18 + 3 \times 4$$

opposites	**order of operations**
Two numbers that are the same distance from 0 on a number line, but on opposite sides of 0 −3 and 3 are opposites.	The order in which to perform operations when evaluating expressions with more than one operation To evaluate $5 + 2 \times 3,$ you perform the multiplication before the addition. $$5 + 2 \times 3 = 5 + 6 = 11$$
ordered pair	**origin**
A pair of numbers (x, y) used to locate a point in a coordinate plane; The first number is the x-coordinate, and the second number is the y-coordinate. The x-coordinate of the point $(-2, 1)$ is $-2,$ and the y-coordinate is 1.	The point, represented by the ordered pair $(0, 0),$ where the horizontal and vertical number lines intersect in a coordinate plane *See coordinate plane.*
outlier	**parallel (lines)**
A data value that is much greater or much less than the other values In the data set 23, 42, 33, 117, 36, and 40, the outlier is 117.	Two lines in the same plane that do not intersect 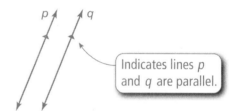
parallelogram	**percent**
A quadrilateral with two pairs of parallel sides 	A part-to-whole ratio where the whole is 100 $$37\% = 37 \text{ out of } 100 = \frac{37}{100}$$

percent of change	**percent of decrease**
The percent that a quantity changes from the original amount	The percent of change when the original amount decreases
percent of change $= \dfrac{\text{amount of change}}{\text{original amount}}$	percent of decrease $= \dfrac{\text{original amount} - \text{new amount}}{\text{original amount}}$
The percent of change from 20 to 25 is: $\dfrac{25 - 20}{20} = \dfrac{5}{20} = 25\%$	The price of a shirt decreases from \$20 to \$10. The percent of decrease is $\dfrac{20 - 10}{20}$, or 50%.

percent error	**percent of increase**
The percent that an estimated quantity differs from the actual amount	The percent of change when the original amount increases
percent error $= \dfrac{\text{amount of error}}{\text{actual amount}}$	percent of increase $= \dfrac{\text{new amount} - \text{original amount}}{\text{original amount}}$
Estimated length: 16 feet Actual length: 21	The price of a shirt increases from \$20 to \$30.
Percent error: $\dfrac{21 - 16}{21}$, or 23.8%	The percent of increase is $\dfrac{30 - 20}{20}$, or 50%.

perfect square	**plane**
The square of a whole number	A flat surface that extends without end in all directions
Because $7^2 = 49$, 49 is a perfect square.	

polygon	**polyhedron**
A closed figure in a plane that is made up of three or more line segments that intersect only at their endpoints	A solid whose faces are all polygons

positive numbers	**power**
Numbers that are greater than 0 $\qquad$ 0.5, 2, 100	A product of repeated factors base $\rightarrow$ exponent $\underbrace{3^4}_{power} = \underbrace{3 \bullet 3 \bullet 3 \bullet 3}_{3 \text{ is used as a factor 4 times.}}$
prime factorization	**prime number**
A composite number written as the product of its prime factors $\qquad 60 = 2 \times 2 \times 3 \times 5$	A whole number greater than 1 with exactly two factors, 1 and itself $\qquad$ 2, 3, 5, 7, 11, 13, 17, 19, 23, 29, 31, ...
principal	**prism**
An amount of money borrowed or deposited You deposit \$200 in an account that earns 4% simple interest per year. The principal is \$200.	A polyhedron that has two parallel, identical bases; The lateral faces are parallelograms. 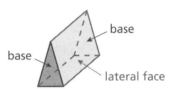
product	**proportion**
The result when two or more numbers are multiplied The product of 4 and 3 is 4×3, or 12.	An equation stating that two ratios are equivalent $\qquad \dfrac{3}{4} = \dfrac{12}{16}$

proportional	**pyramid**
Two quantities that form a proportion are proportional. Because $\dfrac{3}{4}$ and $\dfrac{12}{16}$ form a proportion, $\dfrac{3}{4}$ and $\dfrac{12}{16}$ are proportional.	A polyhedron that has one base; The lateral faces are triangles.
quadrants	**quadrilateral**
The four regions created by the intersection of the horizontal and vertical number lines in a coordinate plane *See coordinate plane.*	A polygon with four sides
quartiles	**quotient**
The quartiles of a data set divide the data into four equal parts. median (second quartile) = 12 lower half upper half 3 4 8 16 21 26 first quartile, Q_1 third quartile, Q_3	The result of a division The quotient of 10 and 5 is $10 \div 5$, or 2.
range (of a data set)	**rate**
The difference between the greatest value and the least value of a data set The range of the data set 12, 16, 18, 22, 27, 35 is $35 - 12 = 23$.	A ratio of two quantities using different units You read 3 books every 2 weeks.

ratio	**ratio table**
A comparison of two quantities; The ratio of a to b can be written as $a : b$. Ratios can be part-to-part, part-to-whole, or whole-to-part comparisons. <p style="text-align:center">$4 : 1$</p>	A table used to find and organize equivalent ratios 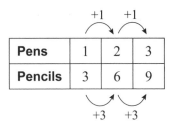
rational number	**reciprocals**
A number that can be written as $\dfrac{a}{b}$ where a and b are integers and $b \neq 0$ $3 = \dfrac{3}{1}, \qquad -\dfrac{2}{5} = \dfrac{-2}{5}$ $0.25 = \dfrac{1}{4}, \qquad 1\dfrac{1}{3} = \dfrac{4}{3}$	Two numbers whose product is 1 Because $\dfrac{4}{5} \times \dfrac{5}{4} = 1$, $\dfrac{4}{5}$ and $\dfrac{5}{4}$ are reciprocals.
rectangle	**repeating decimal**
A parallelogram with four right angles 	A decimal that has a pattern that repeats <p style="text-align:center">$0.555... = 0.\overline{5}$</p><p style="text-align:center">$1.727272... = 1.\overline{72}$</p>
right angle	**simple interest**
An angle whose measure is $90°$ 	Money paid or earned only on the principal You put \$200 into an account. The account earns 5% simple interest per year. The interest earned after 3 years is $\$200 \times 0.05 \times 3$, or \$30. The account balance is $\$200 + \$30 = \$230$ after 3 years.

simplest form (of an algebraic expression)	**skewed left**
An algebraic expression is in simplest form when it has no like terms and no parentheses. $$6a + 9a^2, \; 3t + 5$$	The distribution of a data set is skewed left when the "tail" of the graph extends to the left and most of the data are on the right.
skewed right	**slope**
The distribution of a data set is skewed right when the "tail" of the graph extends to the right and most of the data are on the left. 	A ratio of the change in y (vertical change) to the change in x (horizontal change) between any two points on a line; It is a measure of the steepness of a line. $$\text{slope} = \frac{\text{change in } y}{\text{change in } x}$$
solid	**solution (of an equation)**
A three-dimensional figure that encloses a space 	A value that makes an equation true 6 is the solution of the equation $x - 4 = 2$.
solution of an equation in two variables	**solution of an inequality**
An ordered pair that makes an equation in two variables true $(3, 4)$ is a solution of the equation $y = x + 1$.	A value that makes an inequality true A solution of the inequality $x + 3 > 9$ is $x = 12$.

solution set	**square**
The set of all solutions of an inequality	A parallelogram with four sides that have the same length and four right angles

square(d)	**square units**
A number squared is the number raised to an exponent of 2. 5 squared means 5^2, or 25.	The units are measured in square feet $\left(\text{ft}^2\right)$, square meters $\left(\text{m}^2\right)$.

statistical question	**statistics**
A question for which you do not expect to get a single answer "What is the daily high temperature in August?"	The science of collecting, organizing, analyzing, and interpreting data

stem	**stem-and-leaf plot**
Digit or digits on the left of the stem-and-leaf plot *See stem-and-leaf plot.*	A type of data display that uses the digits of data values to organize a data set; Each data value is broken into a stem (digit or digits on the left) and a leaf (digit or digits on the right).

Test Scores

Stem	Leaf
6	6
7	2 7
8	1 1 3 4 4 6 8 8
9	0 0 0 2 7 8
10	0

Key: 9 | 4 = 94 points

Subtraction Property of Equality

When you subtract the same number from each side of an equation, the two sides remain equal.

$$\begin{array}{rcr} x + 4 &=& 5 \\ -4 & & -4 \\ \hline x &=& 1 \end{array}$$

Subtraction Property of Inequality

When you subtract the same number from each side of an inequality, the inequality remains true.

$$\begin{array}{rcr} x + 4 &>& 5 \\ -4 & & -4 \\ \hline x &>& 1 \end{array}$$

---

sum

The result when two or more numbers are added

The sum of 4 and 3 is $4 + 3$, or 7.

surface area

The sum of the areas of all the faces of a solid

$$\begin{aligned} S &= 15 + 15 + 18 + 18 + 30 + 30 \\ &= 126 \text{ in.}^2 \end{aligned}$$

6 in.

5 in.

3 in.

---

symmetric (distribution)

The distribution of a data set is symmetric when the left side of the graph is a mirror image of the right side of the graph.

terminating decimal

A decimal that ends

$$1.5,\ 2.58,\ -5.605$$

---

terms (of an algebraic expression)

The parts of an algebraic expression

The terms of $4x + 7$ are $4x$ and 7.

third quartile (Q_3)

The median of the upper half of a data set

See quartiles.

---

three-dimensional figure	**trapezoid**
A figure that has length, width, and depth 	A quadrilateral with exactly one pair of parallel sides

triangle	**two-dimensional figure**
A polygon with three sides 	A figure that has only length and width

unit analysis	**unit rate**
A process used to decide which conversion factor will produce the appropriate units $$36 \ \cancel{qt} \ \bullet \ \frac{1 \ gal}{4 \ \cancel{qt}} = 9 \ gal$$	A rate that compares a quantity to one unit of another quantity The speed limit is 65 miles per hour.

U.S. customary system	**variable**
System of measurement that contains units for length, capacity, and weight inches, feet, quarts, gallons, ounces, pounds	A symbol that represents one or more numbers x is a variable in $2x + 1$.

Venn diagram

A diagram that uses circles to describe relationships between two or more sets

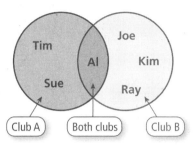

vertex (of a solid)

A point where three or more edges intersect

See face.

volume

A measure of the amount of space that a three-dimensional figure occupies; Volume is measured in cubic units such as cubic feet (ft^3) or cubic meters (m^3).

$V = \ell wh = 12(3)(4) = 144 \text{ ft}^3$

whole numbers

The numbers $0, 1, 2, 3, 4, \ldots$

x-axis

The horizontal number line in a coordinate plane

See coordinate plane.

x-coordinate

The first coordinate in an ordered pair, which indicates how many units to move to the left or right from the origin

In the ordered pair $(3, 5)$, the x-coordinate is 3.

y-axis

The vertical number line in a coordinate plane

See coordinate plane.

y-coordinate

The second coordinate in an ordered pair, which indicates how many units to move up or down from the origin

In the ordered pair $(3, 5)$, the y-coordinate is 5.

Photo Credits

7 Danomyte/Shutterstock.com; **131** NASA;
141 *left* ©iStockphoto.com/james steidl; *right*
U.S. Navy photo by Photographers Mate 2nd Class
Michael Sandberg; **320** Jean Thompson; **327** Baldwin
Online: Children's Literature Project at
www.mainlesson.com

Cartoon Illustrations Tyler Stout

Cover Image Pavelk/Shutterstock.com,
©Dwight Smith/Dreamstime.com,
valdis torms/Shutterstock.com

*Available at *BigIdeasMath.com.*

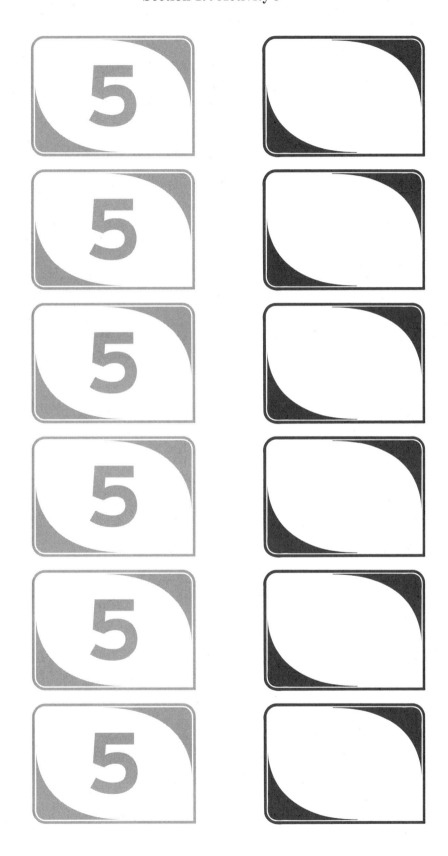

*Available at *BigIdeasMath.com.*

Base Ten Blocks*

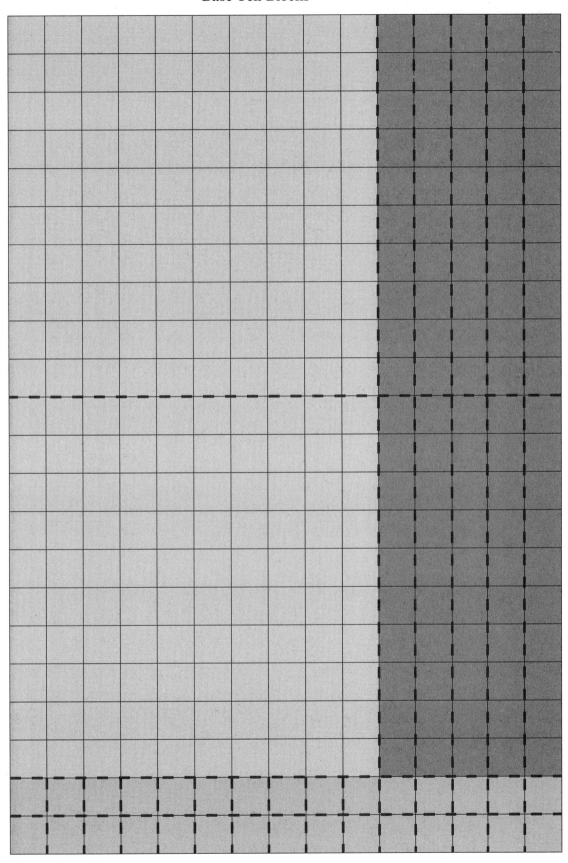

*Available at *BigIdeasMath.com.*

Base Ten Blocks*

*Available at *BigIdeasMath.com*.

*Available at *BigIdeasMath.com.*

*Available at *BigIdeasMath.com*.

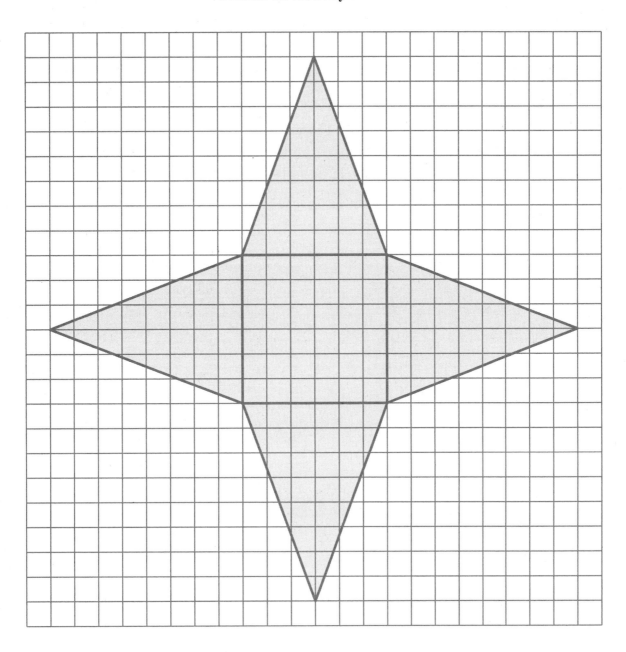

Math Card War – Chapter 12 Section 1*

1.25	-0.6	$-\dfrac{3}{4}$	$\dfrac{3}{10}$	$-\dfrac{3}{2}$
0.75	$\dfrac{19}{10}$	-1.2	$\dfrac{8}{5}$	$\dfrac{3}{20}$
$\dfrac{3}{4}$	-0.4	-0.75	1.9	1.5
0.6	$-\dfrac{2}{5}$	1.6	$-\dfrac{3}{10}$	$\dfrac{6}{5}$

*Available at *BigIdeasMath.com*.

-0.3	0.3	$-\dfrac{19}{10}$	$-\dfrac{8}{5}$
-1.6	1.2	$-\dfrac{3}{20}$	0.4
$\dfrac{3}{5}$	$-\dfrac{3}{5}$	$\dfrac{2}{5}$	-1.25
$\dfrac{5}{4}$	$-\dfrac{6}{5}$	-1.9	$\dfrac{3}{2}$
-0.15	-1.5	$-\dfrac{5}{4}$	0.15

*Available at *BigIdeasMath.com.*

Math Card War – Chapter 15 Section 2*

0.3	$\dfrac{1}{4}$	0.05	0.125
$\dfrac{3}{10}$	40%	$\dfrac{27}{100}$	$\dfrac{1}{8}$
$\dfrac{1}{3}$	12.5%	5%	$\dfrac{1}{20}$
$\dfrac{3}{4}$	$66\dfrac{2}{3}\%$	30%	0.01
75%	0.75	0	1

*Available at *BigIdeasMath.com*.

0.27	$\frac{2}{3}$	1%	0%
100%	$\frac{1}{100}$	27%	0.666...
0.25	0.04	0.333...	0.005
0.4	0.5%	$\frac{2}{5}$	$\frac{1}{200}$
25%	4%	$33\frac{1}{3}\%$	$\frac{1}{25}$

*Available at *BigIdeasMath.com.*

Integer Counters*

Algebra Tiles*

*Available at *BigIdeasMath.com*.